JONES & BARTLETT LEARNING

CDX Automotive

FUNDAMENTALS OF

Automotive Maintenance and Light Repair

STUDENT WORKBOOK

Kirk T. VanGelder

ASE Certified Master Automotive Technician & LI
NATEF Evaluation Team Leader
Certified Automotive Service Instructor
Vancouver, Washington

JONES & BARTLETT
LEARNING

World Headquarters
Jones & Bartlett Learning
5 Wall Street
Burlington, MA 01803
978-443-5000
info@jblearning.com
www.jblearning.com

Jones & Bartlett Learning books and products are available through most bookstores and online booksellers. To contact Jones & Bartlett Learning directly, call 800-832-0034, fax 978-443-8000, or visit our website, www.jblearning.com.

Substantial discounts on bulk quantities of Jones & Bartlett Learning publications are available to corporations, professional associations, and other qualified organizations. For details and specific discount information, contact the special sales department at Jones & Bartlett Learning via the above contact information or send an email to specialsales@jblearning.com.

Production Credits

Chief Executive Officer: Ty Field
President: James Homer
Chief Product Officer: Eduardo Moura
Executive Publisher—CDX and Electrical: Vernon Anthony
Acquisitions Editor—CDX: Ian Andrew
Managing Editor—CDX Automotive: Amanda J. Mitchell
Editorial Assistant: Jamie Dinh
Production Editor: Jessica deMartin
Senior Marketing Manager: Brian Rooney
VP, Manufacturing and Inventory Control: Therese Connell
Composition: Cenveo Publisher Services
Cover Design: Kristin E. Parker
Cover Image: © Mikhail Bakunovich/Shutterstock, Inc.
Printing and Binding: Edwards Brothers Malloy
Cover Printing: Edwards Brothers Malloy

ISBN: 978-1-284-07783-4

6048

Printed in the United States of America
19 18 17 16 15   10 9 8 7 6 5 4 3 2 1

Contents

Careers in Automotive Technology

Chapter Review

The following activities have been designed to help you refresh your knowledge of this chapter. Your instructor may require you to complete some or all of these activities as a regular part of your training program. You are encouraged to complete any activity that your instructor does not assign as a way to enhance your learning.

Matching

Match the following terms with the correct description or example.

A. Automotive Service Excellence (ASE)

B. Brake technician

C. Drivability technician

D. Heavy line technician

E. Lube technician

F. Service consultant/advisor

G. Shop foreman

_____ **1.** Specializes in major engine, transmission, and differential overhaul and repair.

_____ **2.** Carries out scheduled maintenance activities.

_____ **3.** Supervisor who oversees the work of technicians and staff.

_____ **4.** An independent, nonprofit organization dedicated to the improvement of vehicle repair through the testing and certification of automotive professionals.

_____ **5.** Specializes in working on vehicle brake systems.

_____ **6.** Diagnoses and identifies mechanical and electrical faults that affect vehicle performance and emissions.

_____ **7.** A service worker who works with both customers and technicians.

Multiple Choice

Read each item carefully, and then select the best response.

_____ **1.** Who is generally acknowledged to have invented the modern automobile around 1885?

  A. Henry Ford

  B. Karl Benz

  C. Armand Peugeot

  D. Charles Rolls

_____ **2.** Who is a repair shop's first point of contact for customers seeking vehicle repairs?

  A. Shop foreman

  B. Service consultant/advisor

  C. Drivability technician

  D. Service manager

_____ 3. What type of technician might specialize in particular vehicle systems, such as engines, transmissions, or final drives?
   A. Shop foreman
   B. Light line technician
   C. Drivability technician
   D. Heavy line technician

_____ 4. What type of technician diagnoses and repairs faults, replaces or overhauls brake systems, and tests the components of disc, drum, and power brake systems used on all types of vehicles?
   A. Light line technician
   B. Heavy line technician
   C. Brake technician
   D. Drivability technician

_____ 5. What type of technician works with computer-controlled engine management systems to service, identify, and repair faults on electronically controlled vehicle systems such as fuel injection, ignition, antilock braking, cruise control, and automatic transmissions?
   A. Electrical technician
   B. Heavy line technician
   C. Shop foreman
   D. Service manager

_____ 6. What type of technician regularly uses meters, oscilloscopes, circuit wiring diagrams, and solder equipment?
   A. Light line technicians
   B. Transmission specialist
   C. Electrical technicians
   D. Heavy line technicians

_____ 7. Which of the following is a key skill of a service manager?
   A. Communicating
   B. Motivating
   C. Creating positive work environments
   D. All of the above

_____ 8. What types of shops are usually independent and focus on one type of service, such as transmission service, electrical system repair, or emission system diagnosis?
   A. Dealerships
   B. Specialty shops
   C. Franchises
   D. Fleet shops

_____ 9. Which of the following programs receives access to new vehicle technology as well as manufacturer service information to help prepare students for working on today's vehicles and technology?
   A. National Automotive Technicians Education Foundation
   B. Automotive Service Excellence
   C. Automotive Youth Educational Systems
   D. Advanced Engine Performance certification

_____ 10. Technicians who handle refrigerants or work on AC systems are required to have what?
   A. Environmental Protection Agency Section 609 certification
   B. ASE Advanced Engine Performance certification
   C. NATEF refrigerant certification
   D. AYES R134a certification

True/False

If you believe the statement to be more true than false, write the letter "T" in the space provided. If you believe the statement to be more false than true, write the letter "F".

_____ **1.** Today's vehicles are assembled on high-volume production lines, with robots used for many of the assembly processes, including welding seams.

_____ **2.** Heavy line technicians diagnose and replace the mechanical and electrical components of motor vehicles, such as gaskets, belts, hoses, timing belts, water pumps, radiators, alternators, and starters.

_____ **3.** Drivability technicians perform wheel alignments and wheel balancing, and they diagnose and replace faulty steering system components.

_____ **4.** In larger shops, roles may be assigned to separate electrical and drivability technicians, whereas in smaller shops, one technician could perform both roles.

_____ **5.** Electrical technicians test and replace faulty charging system components, starter motors, and related items such as batteries.

_____ **6.** Light line technicians use electronic test equipment, scan tools, pressure transducers, exhaust gas analyzers, lab scopes, meters, and circuit wiring diagrams to locate electrical, fuel, and emission systems faults.

_____ **7.** A service manager's job is to oversee technicians' work in order to ensure that customers receive quality repair work.

_____ **8.** Because dealership technicians are working on the latest vehicles, they are right at the cutting edge of technology.

_____ **9.** The automotive service industry in the United States is generally not subject to licensure requirements.

_____ **10.** The Automotive Youth Educational Systems (AYES) is an independent, nonprofit organization dedicated to the improvement of vehicle repair through the testing and certification of automotive professionals.

Fill in the Blank

Read each item carefully, and then complete the statement by filling in the missing word(s).

1. Henry Ford applied two concepts that helped make the Model T affordable for the masses, _____ and the _____ _____.

2. A(n) _____ _____ changes oil and filters and carries out lubrication, fluid inspection, fluid service, and tire rotations.

3. A(n) _____ _____ diagnoses, repairs, and services steering system components and suspension systems on all types of vehicles.

4. A(n) _____ _____ diagnoses, replaces, maintains, identifies faults with, and repairs electrical wiring and computer-based equipment in vehicles.

5. Electrical technicians use meters, oscilloscopes, test instruments, and circuit wiring diagrams to diagnose _____ _____.

6. A(n) _____ _____ works with computer-controlled engine management systems to service, identify, and repair faults on electronically controlled vehicle systems such as fuel injection, ignition, and automatic transmissions.

7. A(n) _____ _____ may also work on the other components of the drivetrain, including the drive shafts and differentials.

8. A(n) _____ _____ oversees the work of all types of technicians and staff, communicates with customers and external suppliers, and handles the various administrative duties involved with operating a business.

9. _____ are affiliated with a specific vehicle manufacturer.

10. To earn _____ _____, technicians are required to pass one or more ASE certification tests and have 2 years of qualifying work experience as a technician.

Crossword Puzzle

Use the clues to complete the puzzle.

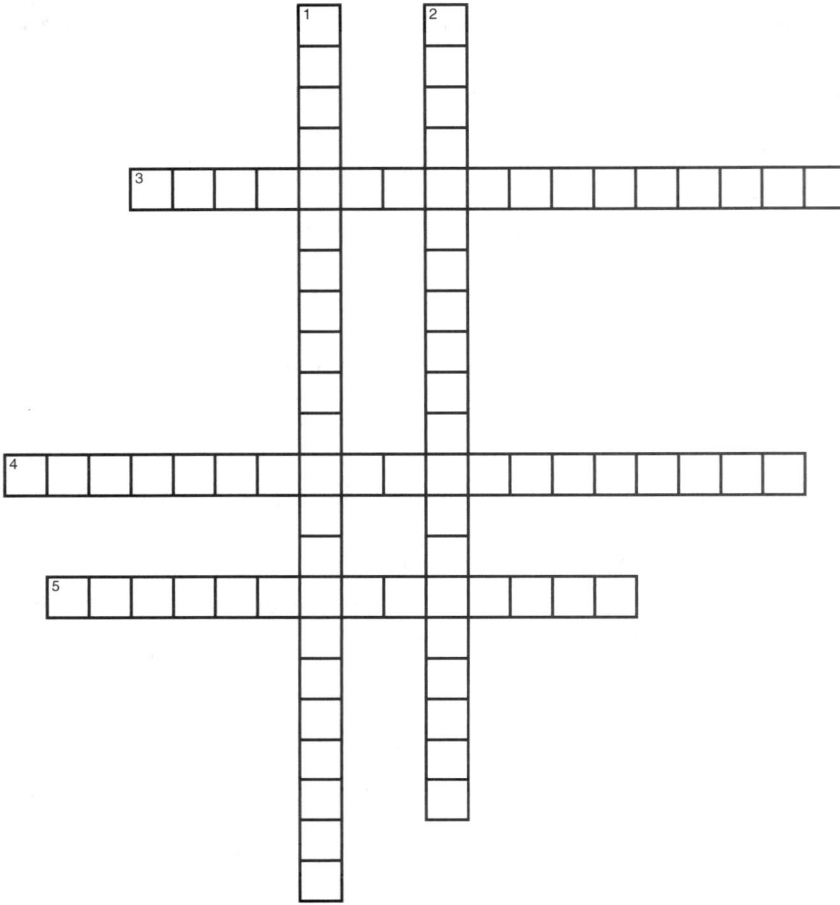

Across

3. A technician who specializes in working on vehicle suspension and steering systems.
4. A technician who diagnoses and replaces the mechanical and electrical components of motor vehicles.
5. The shop supervisor who is responsible for the management of the service department.

Down

1. A technician who diagnoses, overhauls, and repairs transmissions.
2. A technician who diagnoses, replaces, maintains, identifies fault with, and repairs electrical wiring and computer-based equipment in vehicles.

ASE-Type Questions

Read each item carefully, and then select the best response.

_____ 1. Tech A says that newer vehicles require less maintenance compared to older vehicles. Tech B says that service intervals for an older vehicle can be extended if new oils are used. Who is correct?
   A. Tech A
   B. Tech B
   C. Both A and B
   D. Neither A nor B

_____ 2. Tech A says that Henry Ford is credited with the invention of the automobile. Tech B says that Carl Benz is credited with the invention of the automobile. Who is correct?
   A. Tech A
   B. Tech B
   C. Both A and B
   D. Neither A nor B

_____ 3. Tech A says that the production of vehicles today requires a mix of robotic and human assembly to be profitable. Tech B says that most parts on a car are preassembled before they reach the assembly line for higher assembly numbers per day. Who is correct?
   A. Tech A
   B. Tech B
   C. Both A and B
   D. Neither A nor B

_____ 4. Tech A says that the automotive industry is highly technical and only a certain few people will find jobs. Tech B says the automotive industry is wide open with job opportunities for almost every level of skill. Who is correct?
   A. Tech A
   B. Tech B
   C. Both A and B
   D. Neither A nor B

_____ 5. Tech A says that a technician can specialize in different areas based on his or her interest and ability. Tech B says that when a technician specializes in a certain area, he or she will only work on certain vehicle models. Who is correct?
   A. Tech A
   B. Tech B
   C. Both A and B
   D. Neither A nor B

_____ 6. Tech A says that the foreman is the frontline contact for customer relations. Tech B says that the service consultant is the frontline contact for customer relations. Who is correct?
   A. Tech A
   B. Tech B
   C. Both A and B
   D. Neither A nor B

_____ 7. Tech A says that dealership technicians generally have access to manufacturers' training to help prepare them as technicians. Tech B says that an independent shop works on a wide variety of equipment that requires a broad skill level in technicians. Who is correct?
   A. Tech A
   B. Tech B
   C. Both A and B
   D. Neither A nor B

_____ **8.** Tech A says that AYES certifies technicians. Tech B says that ASE certifications can help get you a job. Who is correct?
   A. Tech A
   B. Tech B
   C. Both A and B
   D. Neither A nor B

_____ **9.** Tech A says that the maintenance requirements of a vehicle have not changed since the creation of the automobile. Tech B says that manufacturers are predicting 25,000-mile (40,000-km) intervals. Who is correct?
   A. Tech A
   B. Tech B
   C. Both A and B
   D. Neither A nor B

_____ **10.** Tech A says that a technician can progress to different jobs within the industry. Tech B says that careers in the automotive industry include new car assembly lines. Who is correct?
   A. Tech A
   B. Tech B
   C. Both A and B
   D. Neither A nor B

Introduction to Automotive Technology

CHAPTER 2

Tire Tread: © AbleStock

Chapter Review

The following activities have been designed to help you refresh your knowledge of this chapter. Your instructor may require you to complete some or all of these activities as a regular part of your training program. You are encouraged to complete any activity that your instructor does not assign as a way to enhance your learning.

Matching

Match the following terms with the correct description or example.

A. Differential gear set
B. Four-wheel drive
C. Horizontally opposed engine
D. In-line engine
E. Longitudinal
F. Piston engine
G. Rotary engine
H. Torque converter
I. Unibody design
J. V engine

_____ 1. An engine in which the cylinders are arranged side by side in a single row.

_____ 2. An internal combustion engine that uses cylindrical pistons moving back and forth in a cylinder to extract mechanical energy from chemical energy.

_____ 3. A term used to describe an engine configuration that uses a single bank of cylinders staggered at a shallow 15-degree V.

_____ 4. An engine that uses a triangular rotor turning in a housing instead of conventional pistons.

_____ 5. A term used to describe the front-to-back engine orientation when mounted in the engine compartment.

_____ 6. The arrangement of gears between two axles that allows each axle to spin at its own speed when the vehicle is going around a corner.

_____ 7. A drive train layout in which the engine drive has either two wheels or four wheels depending on which mode is selected by the driver.

_____ 8. A vehicle design that does not use a rigid frame to support the body. The body panels are designed to provide the strength for the vehicle.

_____ 9. A device that is turned by the crankshaft and transmits torque to the input shaft of an automatic transmission.

_____ 10. An engine with two banks of cylinders, 180 degrees apart, on opposite sides of the crankshaft. It is also called a flat engine or a boxer engine.

Multiple Choice

Read each item carefully, and then select the best response.

_____ 1. Which vehicle design has an enclosed body with a maximum of four doors, and a trunk located in the rear of the vehicle accessible from a trunk lid?
A. Coupe
B. Sedan
C. Hatchback
D. Station wagon

_____ **2.** Which type of vehicle acts like both a full-size van and a pickup in that it has a heavier-duty chassis so it can carry heavier loads?
   A. Sedan
   B. Hatchback
   C. Sport utility vehicle
   D. Minivan

_____ **3.** What type of chassis design was first used in aircraft and then spread to automobiles?
   A. Unibody
   B. Body on frame
   C. Dual shell
   D. Steel ladder

_____ **4.** In which type of drivetrain layout are all four wheels driven by the engine all of the time?
   A. Front-wheel drive
   B. Rear-wheel drive
   C. All-wheel drive
   D. Four-wheel drive

_____ **5.** All of the following criteria are used to define the drive train layout, _except_:
   A. Engine position
   B. Transmission type
   C. Engine orientation
   D. Type of drive

_____ **6.** Which of the following engine designs is the most powerful compared to its overall dimensions, but more complicated and expensive than the other engines?
   A. V8
   B. Flat 6
   C. W12
   D. In-line 4

_____ **7.** Which type of engine uses a single bank of cylinders, staggered at a shallow 15-degree V within the bank?
   A. Horizontally opposed
   B. W
   C. V
   D. VR

_____ **8.** Which type of axle uses the engine's torque to turn the wheels (drive the vehicle) and at the same time support the weight of the vehicle?
   A. Live axle
   B. Dead axle
   C. Transaxle
   D. Solid axle

_____ **9.** The twisting force applied to a shaft is known as what?
   A. Play
   B. Torque
   C. Collar
   D. Give

_____ **10.** What designation is used when measuring torque?
   A. Foot-pound
   B. Inch-pound
   C. Newton meter
   D. Any of the above

True/False

If you believe the statement to be more true than false, write the letter "T" in the space provided. If you believe the statement to be more false than true, write the letter "F".

_____ 1. Reducing the number of doors to the passenger compartment makes the vehicle structure more rigid.

_____ 2. In some vehicles, known as roadsters, the roof can be a series of folding steel or fiberglass panels.

_____ 3. A station wagon has an extended roof that goes all the way to the rear of the vehicle. It is similar to a van but not as tall.

_____ 4. Body-on-frame is the term used when a vehicle body is mounted on a rigid frame or chassis.

_____ 5. Some high-performance racing cars today have no chassis at all.

_____ 6. Mechanical energy can be converted into chemical energy in two primary ways: through the operation of an internal combustion engine or through the operation of an electric motor.

_____ 7. The suspension system makes the connection between the steering wheel and the road wheels so the driver can point the vehicle in the intended direction of travel.

_____ 8. A drive train is classified by type, cylinder arrangement, number of cylinders/rotors, and total engine displacement in cubic inches or liters.

_____ 9. Multi-cylinder internal combustion automotive engines are produced in four common configurations.

_____ 10. V engines have two banks of cylinders sitting side by side in a V arrangement sharing a common crankshaft.

_____ 11. Horizontally opposed engines are very powerful for their size, but they do not use conventional pistons that slide back and forth inside a straight cylinder.

_____ 12. The automatic transmission uses a torque converter instead of a clutch.

_____ 13. Part-time 4WD means the vehicle is usually driven in two-wheel drive and switched to full-time when needed by engaging the transfer case.

_____ 14. A transfer case locks the drive shafts together and directs torque through them to both axles.

_____ 15. All transfer cases use a viscous coupling to split the drive between the front and rear wheels.

Fill in the Blank

Read each item carefully, and then complete the statement by filling in the missing word(s).

1. A(n)_____ has only two doors.

2. A(n)_____ is available in three-door and five-door designs.

3. A(n) _____ is an automobile that can convert from having an enclosed top to having an open top by means of a roof that can be removed, retracted, or folded away.

4. A(n) pickup, or _____, carries and tows cargo.

5. A(n) _____ _____ _____ can easily be used to carry out functions that would otherwise require several different vehicles.

6. A(n) _____ is an underlying supporting structure for vehicles—similar to the skeleton of a human—on which additional components are mounted.

7. The _____ design is constructed of a large number of steel sheet metal panels that are precisely formed in presses and spot-welded together into a structural unit.

8. Stored _____ energy is converted to mechanical energy to propel a vehicle down the road.

9. As the pistons move up and down, they rotate the crankshaft, turning the _____ or flex plate, which is bolted to the engine crankshaft.

10. The _____ system evens out the road shocks caused by irregular road surfaces.

11. Manufacturers mount engines in one of two orientations, _____ and _____, depending on which design best fits the vehicle and the rest of the drive train.

12. In a piston engine, the way engine cylinders are arranged is called the engine _____.

13. _____ _____ engines are sometimes referred to as "flat" engines and are commonly found in 4- and 6-cylinder configurations.

14. Axles come in two configurations: _____ axle and _____ axle.

15. A vehicle with a manual transmission uses a _____ to engage and disengage the engine from the transmission.

Labeling

Label the following images with the correct terms.

1. Chassis, engines, and axles:

A. _____

B. _____

C. _____

D. _____

E. _____

2. Manual transmission clutch:

A. _____

B. _____

C. _____

D. _____

E. _____

F. _____

G. _____

H. _____

I. _____

J. _____

K. _____

3. Final drive assembly:

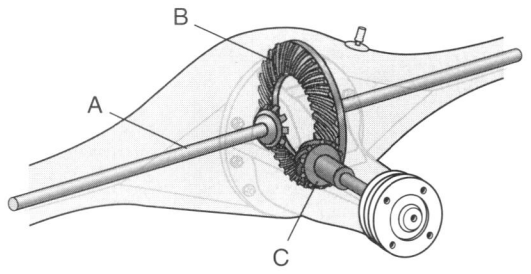

A. _____

B. _____

C. _____

Crossword Puzzle

Use the clues to complete the puzzle.

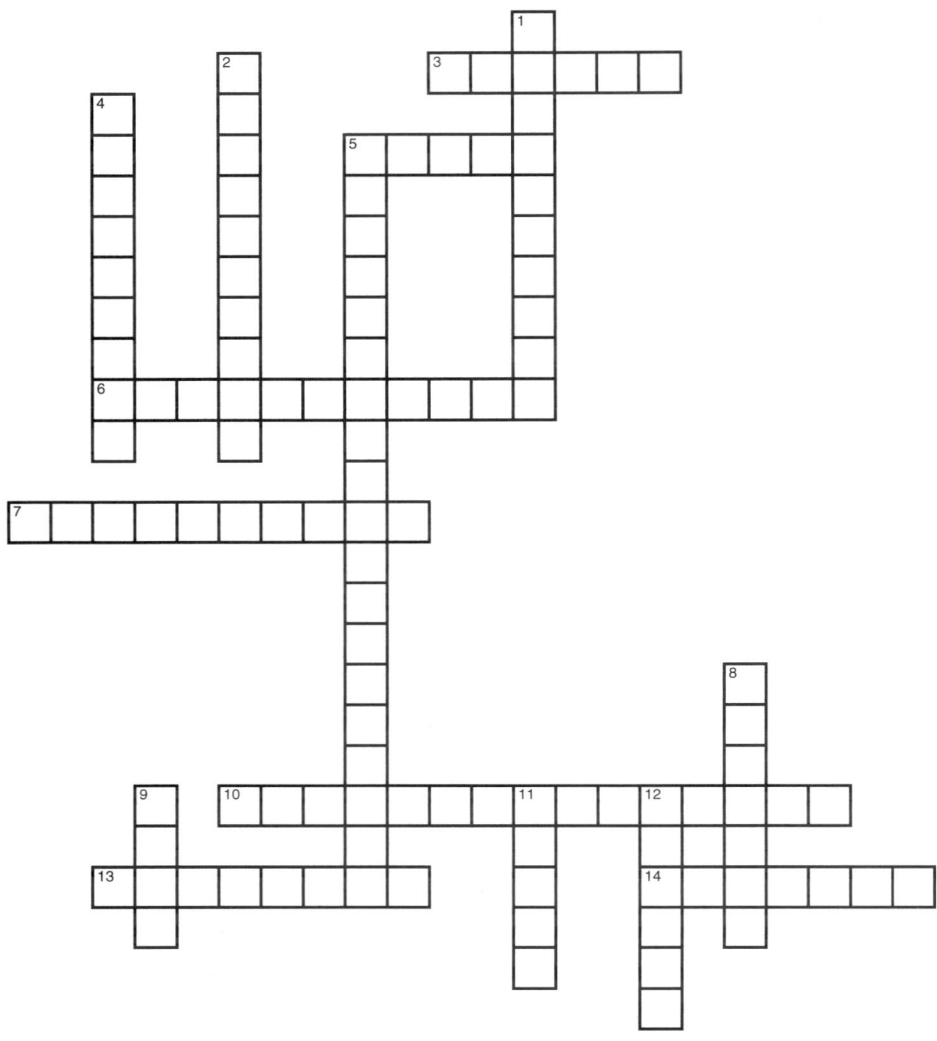

Across

3. Twisting force.

5. A vehicle configuration that has an enclosed body, with a maximum of four doors to allow access to the passenger compartment.

6. A vehicle that converts from having an enclosed top to having an open top by a roof that can be removed, retracted, or folded away.

7. A term used to identify the engine, transmission/transaxle, differential, axles, and wheels.

10. A device that acts like a limited slip clutch.

13. The heavy disc bolted to the rear of the crankshaft that smooths out the power pulses and stores energy from the power stroke for use in keeping the crankshaft rotating through the other three strokes.

14. The main support frame in a vehicle. It includes the running gear, such as suspension, the engine, and the drive train.

Down

1. A term used to describe the side-to-side engine orientation when mounted in the engine compartment.
2. A component that provides a final gear reduction and allows for the difference in speed of each wheel when cornering.
4. A vehicle that has a shared passenger and cargo area; it typically is available in three- and five-door arrangements.
5. A passenger vehicle built on a light truck chassis; it is usually equipped with four-wheel drive and capable of hauling heavier loads than typical passenger vehicles.
8. A lighter-duty van used for carrying six to eight occupants or light cargo.
9. A shaft connected to wheels that transmits the driving torque to the wheels.
11. A two-door vehicle that has seating for two people and may have a small rear seat.
12. A vehicle that carries cargo; it has stronger chassis components and suspension than a sedan.

ASE-Type Questions

Read each item carefully, and then select the best response.

_____ 1. Tech A says that most vehicles today are built with a ladder frame. Tech B says that most vehicles today do not have a frame. Who is correct?
   A. Tech A
   B. Tech B
   C. Both A and B
   D. Neither A nor B

_____ 2. Tech A says that a roadster is a coupe. Tech B says that a station wagon and a hatchback are the same. Who is correct?
   A. Tech A
   B. Tech B
   C. Both A and B
   D. Neither A nor B

_____ 3. Tech A says that gasoline is energy in chemical form. Tech B says that chemical energy is converted to mechanical energy in the combustion process of an engine. Who is correct?
   A. Tech A
   B. Tech B
   C. Both A and B
   D. Neither A nor B

_____ 4. Tech A says that a live axle is an axle that can be steered. Tech B says that a dead axle supports the weight of the vehicle, but doesn't power it. Who is correct?
   A. Tech A
   B. Tech B
   C. Both A and B
   D. Neither A nor B

_____ 5. Tech A says that the gear ratio in the transmission helps the vehicle gain speed. Tech B says that the gear ratio of the final drive can be selected with a gearshift lever. Who is correct?
   A. Tech A
   B. Tech B
   C. Both A and B
   D. Neither A nor B

_____ 6. Tech A says that pushing on the brake pedal stops the vehicle by converting chemical energy into thermal energy. Tech B says that brake pedal pressure is transmitted to the brakes hydraulically. Who is correct?
   A. Tech A
   B. Tech B
   C. Both A and B
   D. Neither A nor B

_____ **7.** Tech A says that one advantage of a mid-engine design is better weight distribution. Tech B says that the rear-engine design is commonly used in four-wheel drive vehicles. Who is correct?
   A. Tech A
   B. Tech B
   C. Both A and B
   D. Neither A nor B

_____ **8.** Tech A says that the purpose of the battery is to charge the vehicle. Tech B says that in-line engines are generally easier to work on than V engines. Who is correct?
   A. Tech A
   B. Tech B
   C. Both A and B
   D. Neither A nor B

_____ **9.** Tech A says that the natural angle of a V6 is 90 degrees. Tech B says that the natural angle of a V10 is 60 degrees. Who is correct?
   A. Tech A
   B. Tech B
   C. Both A and B
   D. Neither A nor B

_____ **10.** Tech A says that the definition of torque is how far the crankshaft twists. Tech B says that torque can relate to tightness of bolts. Who is correct?
   A. Tech A
   B. Tech B
   C. Both A and B
   D. Neither A nor B

Introduction to Automotive Safety

Chapter Review

The following activities have been designed to help you refresh your knowledge of this chapter. Your instructor may require you to complete some or all of these activities as a regular part of your training program. You are encouraged to complete any activity that your instructor does not assign as a way to enhance your learning.

Matching

Match the following terms with the correct description or example.

A. Environmental Protection Agency

B. Hazardous environment

C. Safety data sheet

D. Occupational Safety and Health Administration

E. Personal protective equipment

F. Threshold limit value

_____ 1. Equipment designed to protect the technician, such as safety boots, gloves, clothing, protective eyewear, and hearing protection.

_____ 2. A sheet that provides information about handling, use, and storage of a material that may be hazardous.

_____ 3. The maximum allowable concentration of a given material in the surrounding air.

_____ 4. Government agency created to provide national leadership in occupational safety and health.

_____ 5. A place where hazards exist.

_____ 6. Federal government agency that deals with issues related to environmental safety.

Multiple Choice

Read each item carefully, and then select the best response.

_____ 1. A safe work environment includes which of the following?

A. A well-organized shop layout

B. Good supervision

C. Safety training

D. All of the above

_____ 2. Which federal government agency deals with issues related to environmental safety?

A. SDS

B. EPA

C. OSHA

D. NIOSH

_____ 3. A document that describes the steps required to safely use the vehicle hoist is an example of?

A. Regulation

B. Policy

C. Procedure

D. Compliance

_____ **4.** A _____ can be used to identify hazards and risks within the work environment.
   A. Risk analysis
   B. Procedure
   C. Vulnerability analysis
   D. Threat assessment

_____ **5.** All of the following are components of safety signs, _except_:
   A. Signal word
   B. Background color
   C. Warning light
   D. Pictorial message

_____ **6.** Which signal word indicates a potentially hazardous situation, which, if not avoided, may result in minor or moderate injury?
   A. Danger
   B. Hazard
   C. Warning
   D. Caution

_____ **7.** A droplight that is designed in such a way that the electrical parts can never come into contact with the outer casing of the device is called?
   A. Grounded
   B. Ground fault
   C. Double-insulated
   D. Cordless

_____ **8.** What class of fires involves flammable liquids or gaseous fuels?
   A. Class A
   B. Class B
   C. Class C
   D. Class D

_____ **9.** To operate a fire extinguisher, you should follow which of the following acronyms?
   A. PULL
   B. PASS
   C. PAGE
   D. APES

_____ **10.** The concentration of hazardous material in the air you breathe in your shop must not exceed _____.
   A. The threshold limit value
   B. Ten percent
   C. Shop standards
   D. Fifteen percent

True/False

If you believe the statement to be more true than false, write the letter "T" in the space provided. If you believe the statement to be more false than true, write the letter "F."

_____ **1.** SDS refers to items of safety equipment like safety footwear, gloves, clothing, protective eyewear, and hearing protection.

_____ **2.** There is the possibility of an accident occurring whenever work is undertaken.

_____ **3.** Shop policies and procedures ensure that the shop operates according to OSHA and EPA laws and regulations.

_____ **4.** An OSHA document for the shop that describes how the shop complies with legislation is an example of a procedure.

_____ **5.** There are three standard signal words—danger, warning, and caution.

_____ **6.** Signal words on a warning sign allow the safety message to be conveyed to people who are illiterate or who do not speak the local language.

_____ **7.** Whenever a vehicle's engine is running, toxic gases are emitted from its exhaust.

_____ **8.** An engine fitted with a catalytic converter can be run safely indoors.

_____ **9.** Always keep circuit breaker and electrical panel covers open to provide easy access.

_____ **10.** The danger of a gasoline fire is always present in an automotive shop.

_____ **11.** Fire extinguishers are marked with pictograms depicting the types of fires that the extinguisher is approved to fight.

_____ **12.** Eye wash stations are used to flush the eye with potassium chloride in the event that you get foreign liquid or particles in your eye.

_____ **13.** In the United States it is required that workplaces have an SDS for every chemical that is on site.

_____ **14.** Always use compressed air to blow brake and clutch dust from components and parts before brake service.

_____ **15.** Used oil and fluids will often contain dangerous chemicals and impurities and need to be safely recycled or disposed of in an environmentally friendly way.

Fill in the Blank

Read each item carefully, and then complete the statement by filling in the missing word(s).

1. _____ _____ are a safe way of escaping danger and gathering in a safe place where everyone can be accounted for in the event of an emergency.

2. OSHA stands for the _____ _____ and _____ _____.

3. EPA stands for the _____ _____ _____.

4. It is the _____'s responsibility to know and follow the rules.

5. The signal word _____ indicates a potentially hazardous situation, which, if not avoided, could result in death or serious injury.

6. _____ are used to separate walkways and pedestrian traffic from work areas.

7. _____ _____ is extremely dangerous, as it is odorless and colorless and can build up to toxic levels very quickly in confined spaces.

8. All circuit breakers and fuses should be clearly _____ so you know which circuits and functions they control.

9. Three elements must be present at the same time for a fire to occur: _____, _____, and _____.

10. Class B fire extinguishers are designed for use on flammable liquids or gaseous fuels and are marked with a _____ _____.

11. _____ _____ are designed to smother a small fire and are very useful in putting out a fire on a person.

12. SDS stands for _____ _____ _____.

13. _____ filters can trap very small particles and prevent them from being redistributed into the surrounding air.

14. Coming into frequent or prolonged contact with used _____ _____ can cause dermatitis and other skin disorders, including some forms of cancer.

15. Shop _____ _____ are valuable ways of identifying unsafe equipment, materials, or activities so they can be corrected to prevent accidents or injuries.

Labeling

Label the following images with the correct terms.

To operate a fire extinguisher, follow PASS:

A

B

C

D

A. _____

B. _____

C. _____

D. _____

Crossword Puzzle

Use the clues to complete the puzzle.

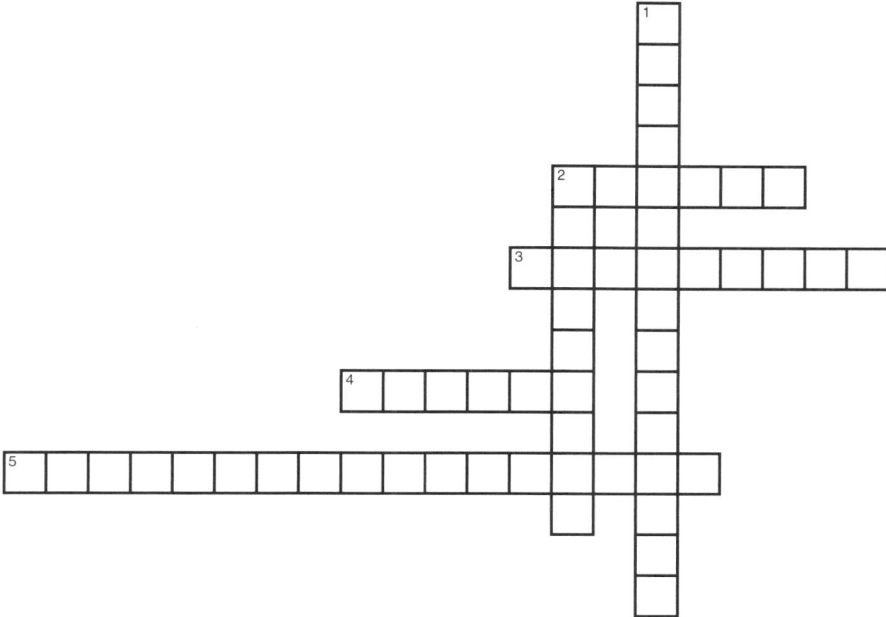

Across

2. A guiding principle that sets the shop direction.

3. Any dust that may contain fine particles that could be harmful to humans or the environment.

4. Anything that could hurt you or someone else.

5. Any material that poses an unreasonable risk of damage or injury to persons, property, or the environment if it is not properly controlled during handling, storage, manufacture, processing, packaging, use and disposal, or transportation.

Down

1. Tools or appliances that are designed in such a way that no single failure can result in a dangerous voltage coming into contact with the outer casing of the device.

2. A list of the steps required to get the same result each time a task or activity is performed.

ASE-Type Questions

Read each item carefully, and then select the best response.

_____ **1.** Tech A says that exposure to solvents may have long-term effects. Tech B says that accidents are almost always avoidable. Who is correct?
  A. Tech A
  B. Tech B
  C. Both A and B
  D. Neither A nor B

_____ **2.** Tech A says that after an accident you should take measures to avoid it in the future. Tech B says that it is OK to block an exit for a shop. Who is correct?
  A. Tech A
  B. Tech B
  C. Both A and B
  D. Neither A nor B

_____ **3.** Tech A says that both OSHA and the EPA can inspect facilities for violations. Tech B says that a shop safety rule does not have to be reviewed once put in place. Who is correct?
   A. Tech A
   B. Tech B
   C. Both A and B
   D. Neither A nor B

_____ **4.** Tech A says that all hazards can be removed from a shop. Tech B says that it is a good practice to disconnect an air gun while inspecting it. Who is correct?
   A. Tech A
   B. Tech B
   C. Both A and B
   D. Neither A nor B

_____ **5.** Tech A says that both caution and danger indicate a potentially hazardous situation. Tech B says that an exhaust extraction hose is not needed if the vehicle is only going to run for a few minutes. Who is correct?
   A. Tech A
   B. Tech B
   C. Both A and B
   D. Neither A nor B

_____ **6.** Tech A says that if you are unsure of what personal protective equipment (PPE) to use to perform a job, you should just use what is nearby. Tech B says that air tools are less likely to shock you than electrically powered tools. Who is correct?
   A. Tech A
   B. Tech B
   C. Both A and B
   D. Neither A nor B

_____ **7.** Tech A says that firefighting equipment includes safety glasses. Tech B says that a class A fire extinguisher can be used to fight an electrical fire only. Who is correct?
   A. Tech A
   B. Tech B
   C. Both A and B
   D. Neither A nor B

_____ **8.** Tech A says that a safety data sheet (SDS) contains information on procedures to repair a vehicle. Tech B says that you only need an SDS if your safety may be in danger. Who is correct?
   A. Tech A
   B. Tech B
   C. Both A and B
   D. Neither A nor B

_____ **9.** Tech A says that a good way to clean dust off brakes is with compressed air. Tech B says that asbestos may be in current auto parts. Who is correct?
   A. Tech A
   B. Tech B
   C. Both A and B
   D. Neither A nor B

_____ **10.** Tech A says that when cleaning brake and clutch components, the wash station should be placed directly under the component. Tech B says that you should follow state and local regulations when disposing of used oil. Who is correct?
   A. Tech A
   B. Tech B
   C. Both A and B
   D. Neither A nor B

Personal Safety

Chapter Review

The following activities have been designed to help you refresh your knowledge of this chapter. Your instructor may require you to complete some or all of these activities as a regular part of your training program. You are encouraged to complete any activity that your instructor does not assign as a way to enhance your learning.

Matching

Match the following terms with the correct description or example.

A. Barrier cream

B. Ear protection

C. Gas welding goggles

D. Heat buildup

E. headgear

F. respirator

G. welding helmet

_____ **1.** Protective gear worn when the sound levels exceed 85 decibels, when working around operating machinery for any period of time, or when the equipment you are using produces loud noise.

_____ **2.** Protective gear designed for gas welding; they provide protection against foreign particles entering the eye and are tinted to reduce the glare of the welding flame.

_____ **3.** A dangerous condition that occurs when a glove can no longer absorb or reflect heat and heat is transferred to the inside of the glove.

_____ **4.** A cream that looks and feels like a moisturizing cream but has a specific formula to provide extra protection from chemicals and oils.

_____ **5.** Protective gear designed for arc welding; it provides protection against foreign particles entering the eye, and the lens is tinted to reduce glare of the welding arc.

_____ **6.** Protective gear that includes items like hairnets, caps, or hard hats.

_____ **7.** Protective gear used to protect the wearer from inhaling harmful dusts or gases.

Multiple Choice

Read each item carefully, and then select the best response.

_____ **1.** Proper footwear provides protection against which of the following?

    A. Chemicals

    B. Cuts

    C. Slips

    D. All of the above

_____ **2.** What type of gloves will protect your hands from burns when welding or handling hot components?

    A. Light-duty

    B. Leather

    C. Chemical

    D. General-purpose cloth

_____ 3. What type of protectant prevents chemicals from being absorbed into your skin and should be applied to your hands before you begin work?
  A. Sunscreen
  B. Moisturizer
  C. Barrier cream
  D. Sanitizer

_____ 4. What type of gloves should be used to protect your hands from exposure to greases and oils?
  A. Chemical
  B. Light-duty
  C. Leather
  D. General-purpose cloth

_____ 5. When using a _____ always make sure the cartridge is the correct type for the contaminant in the atmosphere.
  A. dust mask
  B. welding helmet
  C. respirator
  D. all of the above

_____ 6. The most common type of eye protection is a pair of safety glasses, which must be marked with _____ on the lens and frame.
  A. Z87
  B. PPE
  C. UV
  D. Polarized

_____ 7. What type of protection can be worn instead of a welding mask when using or assisting a person using an oxyacetylene welder?
  A. Respirator
  B. Full face shield
  C. Safety glasses
  D. Gas welding goggles

_____ 8. Which of the following is an example of a mechanical means of ventilation?
  A. Open door
  B. Fume hood
  C. Open window
  D. Both A and C

_____ 9. Before lifting anything, you can reduce the risk of injury by:
  A. Breaking down the load into smaller quantities
  B. Seeking assistance
  C. Using a mechanical device
  D. All of the above

True/False

If you believe the statement to be more true than false, write the letter "T" in the space provided. If you believe the statement to be more false than true, write the letter "F".

_____ 1. PPE includes clothing, shoes, safety glasses, hearing protection, masks, and respirators.

_____ 2. Leather gloves should always be worn when using solvents and cleaners.

_____ 3. Avoid picking up very hot metal with leather gloves because it causes the leather to harden, making it less flexible during use.

_____ 4. Light-duty gloves are designed for use in cold temperatures, particularly during winter, so that cold tools do not stick to your skin.

_____ 5. If a proper barrier cream is not available, use a standard moisturizer as a suitable replacement.

_____ 6. Ear protection that covers the entire outer ear usually has higher noise-reduction ratings than in-the-ear type.

_____ 7. A disposable dust mask should not be used if chemicals, such as paint solvents, are present in the atmosphere.

_____ 8. When grinding, you should wear a pair of safety glasses underneath your face shield for added protection.

_____ 9. Tinted safety glasses are not designed to be worn outside in bright sunlight conditions.

_____ 10. Ultraviolet radiation can burn your skin like a sunburn.

_____ 11. Safety goggles provide much the same eye protection as safety glasses but with added protection against harmful chemicals that may splash up behind the lenses of glasses.

_____ 12. Always remove watches, rings, and jewelry before starting work.

Fill in the Blank

Read each item carefully, and then complete the statement by filling in the missing word(s).

1. It is a good idea to keep a spare set of _____ _____ in the workshop in case a toxic or corrosive fluid is spilled on the ones you are wearing.

2. Your _____ _____ can protect you from bumping your head on a vehicle when the vehicle is raised on a hoist.

3. Never use _____ such as gasoline or kerosene to clean your hands.

4. There are two types of breathing devices: disposable _____ _____ and _____.

5. A(n) _____ has removable cartridges that can be changed according to the type of contaminant being filtered.

6. The light from a welding arc is very bright and contains high levels of _____ radiation.

7. It is necessary to use a full _____ _____ when using solvents and cleaners, epoxies, and resins or when working on a battery.

8. You should always think _____ _____ and then act safely.

9. In high-exposure situations, such as vehicles running in the shop, a mechanical means of _____ is required.

10. Good _____ is about always making sure the shop and your work surroundings are neat and kept in good order.

ASE-Type Questions

Read each item carefully, and then select the best response.

_____ 1. Tech A says that personal protective equipment (PPE) does not include clothing. Tech B says that the PPE used should be based on the task you are performing. Who is correct?
   A. Tech A
   B. Tech B
   C. Both A and B
   D. Neither A nor B

_____ 2. Tech A says that protective clothing that is not in good condition should be replaced. Tech B says that safety glasses are adequate to protect your eyes regardless of the activity. Who is correct?
   A. Tech A
   B. Tech B
   C. Both A and B
   D. Neither A nor B

_____ 3. Tech A says that appropriate work clothes include loose-fitting clothing. Tech B says that you should always wear cuffed pants when working in a shop. Who is correct?
   A. Tech A
   B. Tech B
   C. Both A and B
   D. Neither A nor B

_____ **4.** Tech A says that proper footwear may include both leather- and steel-toed shoes. Tech B says that leather-soled shoes provide slip resistance. Who is correct?
   A. Tech A
   B. Tech B
   C. Both A and B
   D. Neither A nor B

_____ **5.** Tech A says that a hat can help keep your hair clean when working on a vehicle. Tech B says that chemical gloves may be used when working with solvent. Who is correct?
   A. Tech A
   B. Tech B
   C. Both A and B
   D. Neither A nor B

_____ **6.** Tech A says that you should only wear gloves when it is absolutely necessary. Tech B says that leather gloves are used to pick up very hot pieces of metal. Who is correct?
   A. Tech A
   B. Tech B
   C. Both A and B
   D. Neither A nor B

_____ **7.** Tech A says that barrier creams are used to make cleaning your hands easier. Tech B says that hearing protection only needs to be worn by people operating loud equipment. Who is correct?
   A. Tech A
   B. Tech B
   C. Both A and B
   D. Neither A nor B

_____ **8.** Tech A says that dust masks should be used when painting. Tech B says that a respirator should be used when the TLV for a chemical is exceeded. Who is correct?
   A. Tech A
   B. Tech B
   C. Both A and B
   D. Neither A nor B

_____ **9.** Tech A says that tinted safety glasses can be worn when working outside. Tech B says that welding can cause a sunburn. Who is correct?
   A. Tech A
   B. Tech B
   C. Both A and B
   D. Neither A nor B

_____ **10.** Tech A says that you should put tools away when done using them. Tech B says that there is always a risk for injury whenever you lift something. Who is correct?
   A. Tech A
   B. Tech B
   C. Both A and B
   D. Neither A nor B

Vehicle, Customer, and Service Information

Chapter Review

The following activities have been designed to help you refresh your knowledge of this chapter. Your instructor may require you to complete some or all of these activities as a regular part of your training program. You are encouraged to complete any activity that your instructor does not assign as a way to enhance your learning.

Matching

Match the following terms with the correct description or example.

A. Belt routing label
B. Lemon law buyback
C. Refrigerant label
D. Service campaign and recall
E. Shop or service manual

F. Technical Service Bulletin (TSB)
G. Vehicle Emission Control Information (VECI) label
H. Title history
I. Vehicle Identification Number (VIN)
J. Vehicle Safety Certification (VSC) label

_____ **1.** A detailed account of a vehicle's past.

_____ **2.** A label that lists the type and total capacity of refrigerant that is installed in the air conditioning system.

_____ **3.** Information issued by manufacturers to alert technicians of unexpected problems or changes to repair procedures.

_____ **4.** A label that lists a diagram of the serpentine belt routing for the engine accessories.

_____ **5.** A consumer protection law used in some states to identify a new vehicle that has undergone several unsuccessful attempts to repair the same fault.

_____ **6.** A unique serial number that is assigned to each vehicle produced.

_____ **7.** A label certifying that the vehicle meets the Federal Motor Vehicle Safety, Bumper, and Theft Prevention Standards in effect at the time of manufacture.

_____ **8.** A corrective measure conducted by manufacturers when a safety issue is discovered with a particular vehicle.

_____ **9.** Manufacturer's or after-market information on the repair and service of vehicles.

_____ **10.** A label used by technicians to identify engine and emission control information for the vehicle.

Multiple Choice

Read each item carefully, and then select the best response.

_____ **1.** What type of information publication comes in two types—factory and aftermarket?
  A. Owner's manual
  B. Service manual
  C. Technical service bulletin
  D. Labor guide

_____ **2.** Who pays for the costs associated with a mandatory recall?
  A. Consumer
  B. Repair shop
  C. Manufacturer
  D. Insurance company

_____ 3. If a customer wants to know how much it will cost to replace a leaking intake manifold gasket on a particular vehicle, a technician can look up this procedure in the _____ to help estimate the cost.
   A. owner's manual
   B. labor guide
   C. service manual
   D. service information program

_____ 4. When determining labor costs, every tenth of an hour equal _____ minutes.
   A. 4
   B. 5
   C. 6
   D. 7

_____ 5. Initial information on a(n) _____ includes customer and vehicle details, along with a brief description of the customer's complaint(s).
   A. repair order
   B. service bulletin
   C. insurance claim
   D. recall notice

_____ 6. Which of the following is *not* required to determine the total cost of service?
   A. Labor costs
   B. Tax amounts
   C. Cost of gas and consumables used to service the vehicle
   D. Odometer reading

_____ 7. The _____ can provide potential new owners of used vehicles an indication of how well the vehicle was maintained.
   A. service history
   B. owner's manual
   C. repair order
   D. service manual

_____ 8. The vehicle identification number is designed for what type of motor vehicle?
   A. Trucks
   B. Motorcycles
   C. Cars
   D. All of the above

_____ 9. The first digit of a North American vehicle identification number indicates?
   A. Manufacturer
   B. Country of origin
   C. Model
   D. Body type

_____ 10. Which of the three Cs stands for understanding the reason that there is a fault?
   A. Connect
   B. Check
   C. Cause
   D. Control

True/False

If you believe the statement to be more true than false, write the letter "T" in the space provided. If you believe the statement to be more false than true, write the letter "F".

_____ 1. Usually a factory service manual is specific to one year and make of vehicle.

_____ 2. The information found in shop manuals provides a systematic procedure and identifies special tools, safety precautions, and specifications relevant to the task.

_____ 3. To use a technical service bulletin, you need to have a basic understanding of how to start and use a computer.

_____ 4. A typical owner's manual contains step-by-step procedures and diagrams on how to identify if there is a fault and perform an effective repair.

_____ **5.** Online versions of labor guides can be updated as new models of vehicles are released or updates are made by the manufacturer.

_____ **6.** If you are replacing the brake pads on a vehicle with wheel locks, the customer should be charged for the extra time it takes to find the lock key and to remove and install the wheel locks.

_____ **7.** A labor guide is a computer application that is used to identify part numbers for vehicle components.

_____ **8.** Most manufacturers store all service history performed in their dealerships.

_____ **9.** The VIN is a 12-character identification code composed of letters and digits.

_____ **10.** The Vehicle Safety Certification label is used by technicians to identify engine and emission control information for the vehicle.

Fill in the Blank

Read each item carefully, and then complete the statement by filling in the missing word(s).

1. A(n) _____ _____ includes an overview of the controls and features of the vehicle; the proper operation, care, and maintenance of the vehicle; owner service procedures; and specifications or technical data.

2. _____ _____ programs are computer applications used to provide technical information for the repair and maintenance of vehicles.

3. _____ can be mandatory and enforced by law or voluntary in order to ensure the safe operation of the vehicle or damage to their business and product image.

4. When working on vehicles that have not been recalled, check to see if a _____ _____ _____ has been issued for that vehicle and type of repair.

5. _____ _____ are essentially electronic versions of a parts manual, available via a CD/DVD, a computer network, or the Internet.

6. A(n) _____ _____ is used by the technician to guide him or her to the problem, and by the customer service staff to create the invoice when the work is completed.

7. The _____ section contains information about the methods of payment, which can be cash, credit card, or account.

8. A(n) _____ _____ tells you that the vehicle has been wrecked and suffered irreparable damage.

9. The _____ _____ _____ label certifies that the vehicle meets the Federal Motor Vehicle Safety, Bumper, and Theft Prevention Standards in effect at the time of manufacture.

10. The 3 Cs are an easy way to learn the fundamental steps in conducting repairs. They stand for _____, _____, and _____.

Labeling

Label the following images with the correct terms.

1. VIN standards:

A.

1	2	3	4	5	6	7	8	9	10	11	12	13	14	15	16	17
Manufacturer Identifier			Vehicle Attributes					Check Character	Model Year	Plant Code	Sequential Number					

B.

1	2	3	4	5	6	7	8	9	10	11	12	13	14	15	16	17
World Manufacturer Identifier (WMI)		Vehicle Descriptor System (VDS)							Vehicle Identifier System (VIS)							

A. _____

B. _____

2. Vehicle information labels:

A. _____

B. _____

C. _____

D. _____

E. _____

Skill Drills

Read each item carefully, and then place each procedure in the correct sequence.

1. Identifying Correct Service Procedures Using a Shop Manual:
   _____ **A.** Locate the correct section that will contain the information you need. The first page of the shop manual is usually a table of contents.
   _____ **B.** Decide what information you need to know about the job and about the vehicle. Make sure you know the make, model, and year of manufacture of the vehicle, and the type and size of the engine. You should also have the vehicle identification number handy.
   _____ **C.** Locate the vehicle specifications by consulting the specifications page in the proper section.
   _____ **D.** Find the appropriate shop manual for the make, model, and year of the vehicle you are working on.
   _____ **E.** Locate the correct section that will contain the information you need.

2. Using a Service Information Program:
   _____ **A.** The search engine will provide a list of possible matches for you to select from. If the initial search does not produce what you are looking for, try changing the search criteria. Keep searching until you find the information.
   _____ **B.** Enter the vehicle identification information into the system in the appropriate places: year, make, model, engine, and possibly VIN.

_____ **C.** Log in to the application using the appropriate username and password.

_____ **D.** Finally, once the general details for the item are displayed, gather the specific information on the specifications or repairs. You may need more than one piece of information.

_____ **E.** Search for the information you require to perform the service or repair.

_____ **F.** If necessary, start the computer and select the service information program.

_____ **G.** Print out or write down the information needed. Put this on a clipboard and take it with you to perform the service or repair.

3. Using a Labor Guide:

_____ **A.** Check for any "combination" time that would need to be added to the base job when a related job is also being completed.

_____ **B.** Find the labor operation either by working your way through the menu tree or by typing a keyword into the search bar.

_____ **C.** Calculate the total time and multiply it by the shop's hourly labor rate. You now have the correct figure to estimate the charge for the particular service.

_____ **D.** Once you locate the labor operation, there are usually two columns that list the time. The first one is "warranty time." The second time listed is the "customer pay" time.

_____ **E.** Decide what specific labor operations you need to locate. Make sure you know the year, make, model, engine, and any other pertinent details of the vehicle.

_____ **F.** Enter the vehicle information into the system.

_____ **G.** Check for any "additional" time. This is extra time needed to deal with situations that occur on a relatively common basis, such as vehicle-installed options that are not common to all vehicles, like wheel locks.

_____ **H.** Log in to the labor estimating system.

4. Locating Parts Information on the Computer:

_____ **A.** Search for the parts you require to conduct the service or repair.

_____ **B.** Gather information on the identified parts, including part numbers, location, availability, and cost.

_____ **C.** Enter the year, make, model, and engine and VIN number information into the system in the appropriate places.

_____ **D.** Log in to the application using the appropriate username and password.

_____ **E.** The search engine will provide a list of possible matches for you to select from. If the initial search does not produce what you are looking for, try changing the search criteria. Keep searching until you find the information.

5. Applying the 3 Cs:

_____ **A.** Retest the vehicle to be sure the fault has been corrected.

_____ **B.** Review the information you collect from the tests. To review effectively, you need to understand how the systems work and interact.

_____ **C.** Identify and document the correction required, including work activities and parts required or used. Make repairs or replace parts to complete the repair.

_____ **D.** Identify and document the root cause of the concern. Research shop manuals and conduct tests to identify the cause of the problem. This may require a number of tests across multiple systems.

_____ **E.** Using the 3 Cs, document the repair process required for a repair order.

_____ **F.** Fill in the required repair order with details of the work conducted.

_____ **G.** Identify and document the concern. This should be on the repair order. Obtain as much information as possible, as this will help you to understand the problem. Identify what the problem is and which vehicle systems are involved.

_____ **H.** Always work safely and use the proper tools and correct personal protective equipment (PPE).

Crossword Puzzle

Use the clues to complete the puzzle.

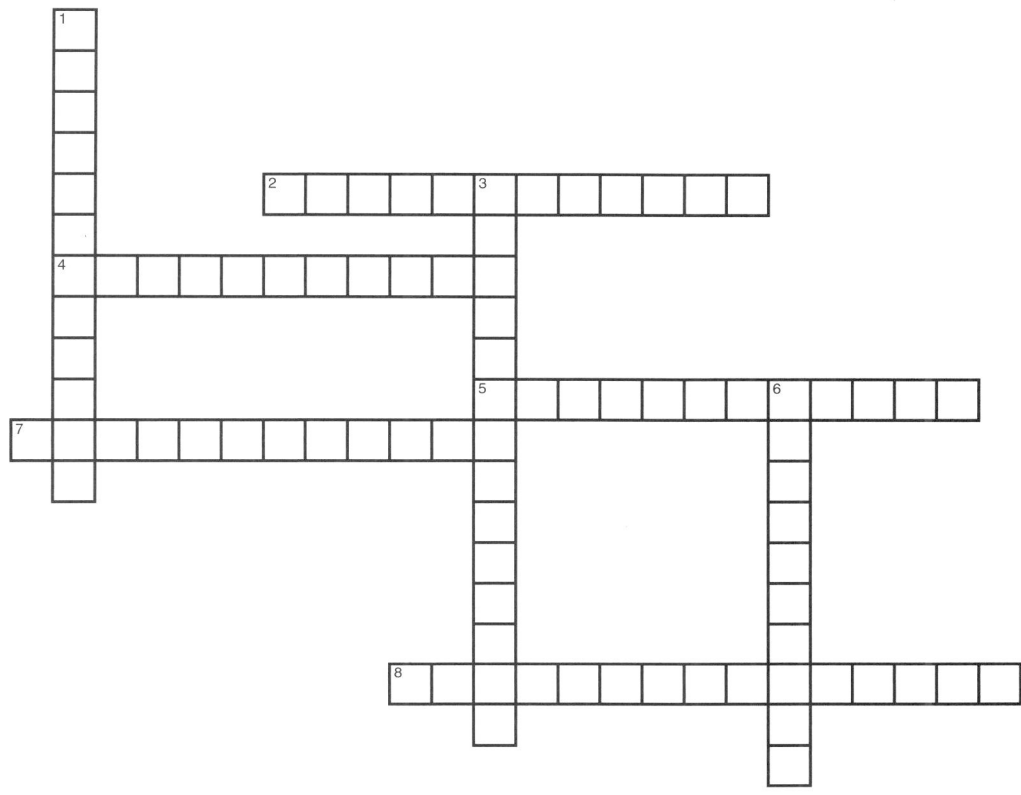

Across

2. An information guide supplied by the manufacturer; it contains basic vehicle operating information.
4. A form used by shops to collect information regarding a vehicle coming in for repair, also referred to as a work order.
5. A label that lists the type of coolant installed in the cooling system.
7. Also called a branded title; a record that a vehicle has been severely damaged or deemed a total loss by an insurance company.
8. The person who serves customers at the parts counters.

Down

1. A computer software program for identifying and ordering replacement vehicle parts.
3. A complete list of all the servicing and repairs that have been performed on a vehicle.
6. A guide that provides information to make estimates for repairs.

ASE-Type Questions

Read each item carefully, and then select the best response.

_____ 1. Tech A says that the owner's manual will have the oil pan drain plug torque information. Tech B says that oil pan capacity information for that specific vehicle will be in the owner's manual. Who is correct?
  A. Tech A
  B. Tech B
  C. Both A and B
  D. Neither A nor B

_____ 2. Tech A says that vehicle security PIN codes can be found in the service manual. Tech B says that vehicle security PIN codes can be found in the owner's manual. Who is correct?
  A. Tech A
  B. Tech B
  C. Both A and B
  D. Neither A nor B

_____ 3. Tech A says that paper service manuals are gone and electronic versions of the service manual are now available. Tech B says that online manuals are more current, as they can be updated periodically. Who is correct?
  A. Tech A
  B. Tech B
  C. Both A and B
  D. Neither A nor B

_____ 4. Tech A says that service information programs are extremely helpful, as the technician can use a laptop at the repair for quick access to the procedure to perform a repair. Tech B says that service information programs allow a technician to know the labor guide and, with the support of the program, perform the task in the time allowed. Who is correct?
  A. Tech A
  B. Tech B
  C. Both A and B
  D. Neither A nor B

_____ 5. Tech A says that TSBs are updates to the owner's manual. Tech B says that TSBs are generally updated information on model changes that do not affect the technician. Who is correct?
  A. Tech A
  B. Tech B
  C. Both A and B
  D. Neither A nor B

_____ 6. Tech A says that labor guides are necessary for the service writer to quote prices for a customer on the repair bill. Tech B says that labor guides are what the customer pays and that the warranty pays more for labor using a different labor guide. Who is correct?
  A. Tech A
  B. Tech B
  C. Both A and B
  D. Neither A nor B

_____ 7. Tech A says that locating the part number needed by the technician requires computer knowledge and mechanical knowledge. Tech B says that anybody can be a parts person. Who is correct?
  A. Tech A
  B. Tech B
  C. Both A and B
  D. Neither A nor B

_____ 8. Tech A says that the repair order is just a piece of paper telling the technician what to do. Tech B says that the repair order is a legal and binding contract between the customer and the repair facility. Who is correct?
  A. Tech A
  B. Tech B
  C. Both A and B
  D. Neither A nor B

_____ **9.** Tech A says that customer authorization for a change in the repair order after initial authorization is the responsibility of the service writer or foreman. Tech B says that changes to the repair order can be dealt with after the repair is complete, as the customer will be satisfied if more repairs are completed. Who is correct?

 A. Tech A

 B. Tech B

 C. Both A and B

 D. Neither A nor B

_____ **10.** Tech A says that the VIN number on a vehicle can help identify which engine is installed in the chassis. Tech B says that most digits in the VIN number can be an identifier for information pertinent to that vehicle. Who is correct?

 A. Tech A

 B. Tech B

 C. Both A and B

 D. Neither A nor B

Tools and Equipment

Chapter Review

The following activities have been designed to help you refresh your knowledge of this chapter. Your instructor may require you to complete some or all of these activities as a regular part of your training program. You are encouraged to complete any activity that your instructor does not assign as a way to enhance your learning.

Matching

Match the following terms with the correct description or example.

A. Bottoming tap
B. Chassis dynamometer
C. Cross-arm
D. Dead blow hammer
E. Dial bore gauge
F. Die stock
G. Flashback arrestor
H. Intermediate tap
I. Micrometer

J. Parallax error
K. Peening
L. Pullers
M. Solder
N. Split ball gauge
O. Telescoping gauge
P. Tensile strength
Q. Vernier calipers
R. Wad punch

_____ **1.** A spring-loaded valve installed on oxyacetylene torches as a safety device to prevent flame from entering the torch hoses.

_____ **2.** A mixture of lead and tin with a low melting point for connecting wires.

_____ **3.** An accurate measuring device for inside bores, usually made with a dial indicator attached to it.

_____ **4.** An accurate measuring device for internal, external, and depth measurements that incorporates fixed and adjustable jaws.

_____ **5.** A thread-cutting tap designed to cut threads to the bottom of a blind hole.

_____ **6.** A gauge that expands and locks to the internal diameter of bores; a caliper or outside micrometer is used to measure its size.

_____ **7.** A type of hammer that has a cushioned head to reduce the amount of head bounce.

_____ **8.** An accurate measuring device for internal and external dimensions.

_____ **9.** In reference to fasteners, the amount of force it takes before a fastener breaks.

_____ **10.** A description for an arm that is set at right angles or 90 degrees to another component.

_____ **11.** A type of punch that is hollow for cutting circular shapes in soft materials such as gaskets.

_____ **12.** One of a series of taps designed to cut an internal thread. Also called a plug tap.

_____ **13.** A term used to describe the action of flattening a rivet through a hammering action.

_____ **14.** A machine with rollers that allows a vehicle to attain road speed and load while sitting still in the shop.

_____ **15.** A generic term to describe hand tools that mechanically assist the removal of bearings, gears, pulleys, and other parts.

_____ **16.** A handle for securely holding dies to cut threads.

_____ **17.** A measuring device used to accurately measure small holes.

_____ **18.** A visual error caused by viewing measurement markers at an incorrect angle.

Multiple Choice

Read each item carefully, and then select the best response.

_____ 1. What device is fitted to compressed air systems to remove the moisture or water from the compressed air that is condensed as a result of compressing air from the atmosphere?
   A. Spit filter
   B. Air drier
   C. Relief valve
   D. Strainer

_____ 2. Which of the following is a type of fastener?
   A. Bolt
   B. Nut
   C. Stud
   D. All of the above

_____ 3. In the metric system, which of the following is measured by the distance between the peaks of threads in millimeters?
   A. Thread pitch
   B. Tensile strength
   C. Torque
   D. Thread count

_____ 4. What type of wrench is also known as a tension wrench?
   A. Open-end wrench
   B. Ratcheting box-end wrench
   C. Torque wrench
   D. Pipe wrench

_____ 5. As long as a bolt is not tightened too much it will return to its original length when loosened; this is called?
   A. Torque
   B. Elasticity
   C. Yield
   D. Play

_____ 6. A(n) _____ is the fastest way to spin a fastener on or off a thread by hand, but it cannot apply much torque to the fastener.
   A. breaker bar
   B. sliding T-handle
   C. speed brace
   D. lug wrench

_____ 7. What type of pliers are used for cutting wire and cotter pins?
   A. Diagonal
   B. Snap ring
   C. Flat-nosed
   D. Needle-nosed

_____ 8. A screw or bolt with a cross-shaped recess requires a(n)_____?
   A. flat blade screwdriver
   B. offset screwdriver
   C. Phillips head screwdriver
   D. Allen wrench

_____ 9. What type of screwdriver fits into spaces where a straight screwdriver cannot and is useful where there is not much room to turn it?
   A. Ratcheting screwdriver
   B. Impact driver
   C. Phillips head screwdriver
   D. Offset screwdriver

_____ **10.** What kind of tools are composed of a strong metal and used as a lever to move, adjust, or pry?
  A. Cold chisels
  B. Pry bars
  C. Drift punches
  D. Speed brace

_____ **11.** What type of file is thinner than other files, comes to a point, and is used for working in narrow slots?
  A. Warding file
  B. Triangular file
  C. Thread file
  D. Square file

_____ **12.** What type of file cleans clogged or distorted threads on bolts and studs?
  A. Triangular file
  B. Warding file
  C. Thread file
  D. Square file

_____ **13.** The name for this type of clamp comes from its shape, it can hold parts together while they are being assembled, drilled, or welded.
  A. J-clamp
  B. C-clamp
  C. D-clamp
  D. K-clamp

_____ **14.** What type of tap narrows at the tip to give it a good start in the hole where the thread is to be cut?
  A. Intermediate tap
  B. Taper tap
  C. Plug tap
  D. Bottoming tap

_____ **15.** What type of tool consists of three main parts: jaws, a cross-arm, and a forcing screw?
  A. Tap and die set
  B. Bench vice
  C. Flaring tool
  D. Gear puller

_____ **16.** What tool is used to measure the gap between a straight edge and the surface being checked for flatness?
  A. Steel rule
  B. Caliper
  C. Feeler gauge
  D. Split ball gauge

_____ **17.** What type of grinder uses discs rather than wheels?
  A. Bench grinder
  B. Angle grinder
  C. Straight grinder
  D. Pedestal grinder

_____ **18.** What tool uses high-pressure to blast small abrasive particles to clean the surface of parts?
  A. Power washer
  B. Angle grinder
  C. Sand blaster
  D. Solvent tank

_____ **19.** What type of torch is occasionally used by technicians to heat, braze, weld, and cut metal?
  A. Ultrasonic
  B. Magnesium
  C. Plasma
  D. Oxyacetylene

_____ **20.** What type of welder has a filler rod that automatically feeds into the welding joint via the hand piece?
  A. Plasma welder
  B. Wire feed welders
  C. Arc welder
  D. Oxyacetylene welder

True/False

If you believe the statement to be more true than false, write the letter "T" in the space provided. If you believe the statement to be more false than true, write the letter "F".

_____ 1. Serious, sometimes fatal, injuries can be caused by compressed air being injected into the body through the skin or into a body opening, such as your mouth or ear.

_____ 2. Thread pitch is a way of defining how much a fastener should be tightened.

_____ 3. The higher the grade number of a fastener, the higher the tensile strength.

_____ 4. Torque wrenches come in various types: beam style, clicker, dial, and electronic.

_____ 5. The open-end wrench fits fully around the head of the bolt or nut and grips each of the six points at the corners just like a socket.

_____ 6. Six- and 12-point sockets fit the heads of hexagonal shaped fasteners.

_____ 7. Arc joint pliers, also called vice grips, are general-purpose pliers used to clamp and hold one or more objects.

_____ 8. Allen wrenches are sometimes called hex keys.

_____ 9. When a large chisel needs a really strong blow, it is time to use a dead blow hammer.

_____ 10. When marks need to be drawn on an object like a steel plate to help locate a hole to be drilled, a drift punch can be used to mark the points so they will not rub off.

_____ 11. A cold chisel gets its name from the fact it is used to cut cold metals, rather than heated metals.

_____ 12. A bottoming tap is used to tap a thread into a hole that does not come out the other side of the material.

_____ 13. Always use a single flare if the tubing is to be used for higher pressures such as in a brake system.

_____ 14. A tubing cutter is more convenient and neater than a saw when cutting pipes and metal tubing.

_____ 15. When soldering, always apply flux to the joint if cored solder is used.

_____ 16. Depth micrometers are used to measure inside dimensions.

_____ 17. Morse taper is a system for securing drill bits to drills.

_____ 18. A solvent tank is a cleaning tank that is filled with a suitable solvent to clean parts by removing oil, grease, dirt, and grime.

_____ 19. Thread repair is used in situations where it is not possible to replace a damaged component.

_____ 20. Batteries are filled with sulphuric acid, so if the hydrogen explodes, the battery case can then rupture and spray everything and everyone nearby with this dangerous and corrosive liquid.

Fill in the Blank

Read each item carefully, and then complete the statement by filling in the missing word(s).

1. _____ / _____ is an umbrella term that describes a set of safety practices and procedures that are intended to reduce the risk of technicians inadvertently using tools, equipment, or materials that have been determined to be unsafe.

2. A(n) _____ _____ is designed to regularly oil an air tool or air equipment so it does not have to be done manually, before or during its use.

3. A(n) _____ is a cylindrical piece of metal with a hexagonal head on one end and a thread cut into the shaft at the other end.

4. Each bolt diameter in the standard system can have one of two thread pitches, _____ or _____.

5. _____-to- _____ means that a fastener is torqued to, or just beyond, its yield point.

6. A(n) _____ wrench has an open-end head on one end and a box-end head on the other end.

7. _____ is a hand tool designed to hold, cut, or compress materials.

8. End cutting pliers, also called _____, have a cutting edge at right angles to their length.

9. A(n) _____ _____ is used when a screw or a bolt is rusted/corroded in place or overtightened and needs a tool that can apply more force.

10. A(n) _____ _____ hammer is designed not to bounce back when it hits something.

11. _____ are used when the head of the hammer is too large to strike the object being hit without causing damage to adjacent parts.

12. A(n) _____ _____ has a hardened, sharpened blade and is designed to remove a gasket without damaging the sealing face of the component.

13. A screw _____ is a device designed to remove screws, studs, or bolts that have broken off in threaded holes.

14. To cut a brand new thread on a blank rod or shaft, a die held in a _____ _____ is used.

15. A measuring _____ is a flexible type of ruler and a common measuring tool.

16. A(n) _____ _____ can measure how round something is.

17. A(n) _____ _____ is used to measure the width of gaps, such as the clearance between valves and rocker arms.

18. The most common air tool in an automotive shop is the air _____ _____.

19. A(n) _____ _____ works by applying a pressurized gas such as air, oxygen, nitrogen, or argon through a nozzle located in the center of a hand piece with an electrode.

20. A(n) _____ _____ incorporates microprocessors to monitor and control the charge rate so a battery receives the correct amount of charge depending on its state of charge.

Labeling

Label the following diagrams with the correct terms.

1. Anatomy of a socket:

A. _____

B. _____

C. _____

D. _____

E. _____

F. _____

2. Components of a flare tool:

A. _____

B. _____

C. _____

D. _____

E. _____

F. _____

G. _____

3. Anatomy of a rivet:

A. _____

B. _____

C. _____

D. _____

4. Air tools:

A. _____

B. _____

C. _____

D. _____

E. _____

5. Oxyacetylene tips:

A. _____

B. _____

C. _____

Crossword Puzzle

Use the clues to complete the puzzle.

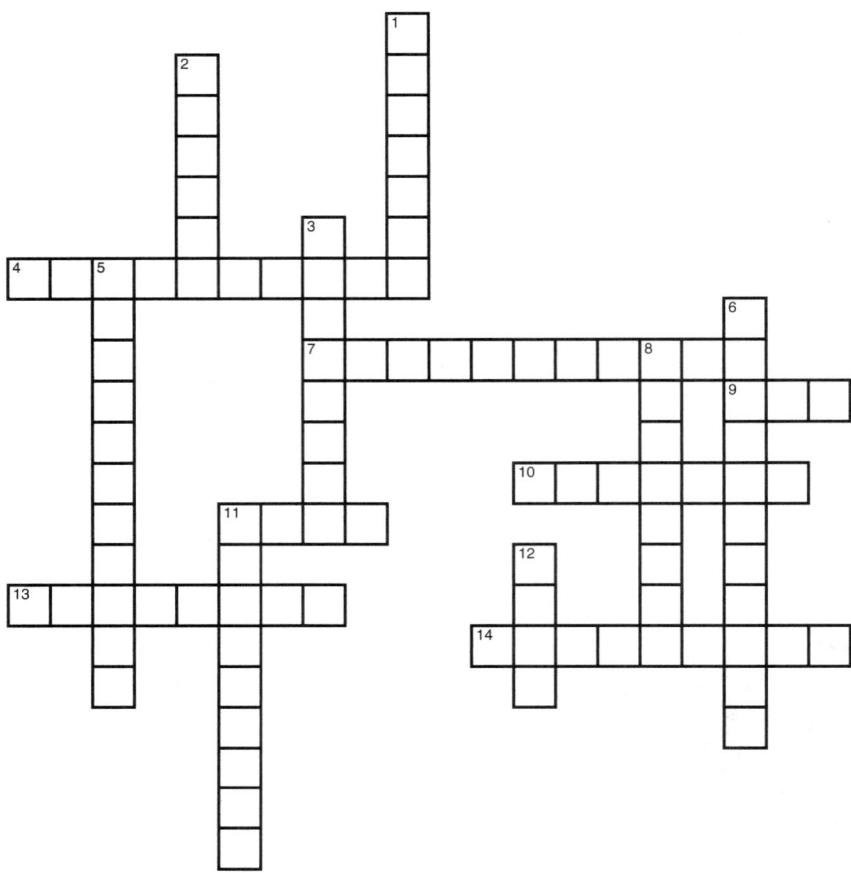

Across

4. The amount of stretch or give a material has.

7. A seal that is made at the end of metal tubing or pipe.

9. A fastener with a hexagonal head and internal threads for screwing on bolts.

10. The shaft of a pop rivet.

11. A type of threaded fastener with a thread on one end and a hexagonal head on the other.

13. A type of punch in various sizes with a straight or parallel shaft.

14. A tool designed to remove wheel lug nuts and commonly shaped like a cross.

Down

1. A device that converts and stores electrical energy through chemical reactions.

2. An enclosed metal tube commonly with 6 or 12 points to remove and install bolts and nuts.

3. A compressed air-powered drill.

5. A type of hexagonal drive mechanism for fasteners.

6. Less sharp than a prick punch, this makes a bigger indentation that centers a drill bit at the point where a hole is required to be drilled.

8. A device fitted to compressed air lines to remove moisture.

11. A device that securely holds material in jaws while it is being worked on.

12. A liquid or paste that protects a soldering or welding joint from oxidization.

ASE-Type Questions

Read each item carefully, and then select the best response.

_____ 1. Tech A says that knowing how to use tools correctly creates a safe working environment. Tech B says that a flare nut wrench is used to loosen very tight bolts and nuts. Who is correct?
   A. Tech A
   B. Tech B
   C. Both A and B
   D. Neither A nor B

_____ 2. Tech A says that lockout/tagout is a safety procedure for a safe working environment. Tech B says that a soldering iron is used to weld, cut, and braze steel components. Who is correct?
   A. Tech A
   B. Tech B
   C. Both A and B
   D. Neither A nor B

_____ 3. Tech A says that torque wrenches need to be calibrated periodically to ensure proper torque values. Tech B says that when a bolt is torqued the bolt stretches beyond its yield point and the bolt has to be replaced when removed. Who is correct?
   A. Tech A
   B. Tech B
   C. Both A and B
   D. Neither A nor B

_____ 4. Tech A says that head bolts are tightened past their yield point. Tech B says that head bolts are torqued then tightened to their yield point. Who is correct?
   A. Tech A
   B. Tech B
   C. Both A and B
   D. Neither A nor B

_____ 5. Tech A says that a box-end wrench is more likely to round the head of a bolt than an open-end wrench. Tech B says that 6-point sockets and wrenches have more surface area on the bolt and will hold more firmly when removing and tightening. Who is correct?
   A. Tech A
   B. Tech B
   C. Both A and B
   D. Neither A nor B

_____ 6. Tech A says that it is usually better to pull a wrench to tighten or loosen a bolt. Tech B says that pushing a wrench will protect your knuckles if the wrench slips. Who is correct?
   A. Tech A
   B. Tech B
   C. Both A and B
   D. Neither A nor B

_____ 7. Tech A says that when jump starting a vehicle, a spark typically occurs when making the last jumper cable connection. Tech B says that all four of the jumper cable connections should be made at the battery terminals. Who is correct?
   A. Tech A
   B. Tech B
   C. Both A and B
   D. Neither A nor B

_____ 8. Tech A says that a dead blow hammer reduces rebound of the hammer. Tech B says that a dead blow hammer should be used to cut the head of a bolt off with a chisel. Who is correct?
   A. Tech A
   B. Tech B
   C. Both A and B
   D. Neither A nor B

_____ **9.** Tech A says that gaskets can be removed quickly and safely with a portable grinder as long as the grinding wheel isn't too coarse. Tech B says that extreme care must be used when removing a gasket on an aluminum surface. Who is correct?

    A. Tech A

    B. Tech B

    C. Both A and B

    D. Neither A nor B

_____ **10.** Tech A says that when using a file, apply pressure to the file in the direction of the cut and no pressure when pulling the file back. Tech B says that file cards are used to file uneven surfaces. Who is correct?

    A. Tech A

    B. Tech B

    C. Both A and B

    D. Neither A nor B

Vehicle Protection and Jack and Lift Safety

Chapter Review

The following activities have been designed to help you refresh your knowledge of this chapter. Your instructor may require you to complete some or all of these activities as a regular part of your training program. You are encouraged to complete any activity that your instructor does not assign as a way to enhance your learning.

Matching

Match the following terms with the correct description or example.

A. Four-post hoist
B. Hydraulic jack
C. Pneumatic jack
D. Safe working load
E. Single-post hoist
F. Two-post hoist

_____ 1. A type of vehicle jack that uses compressed gas or air to lift a vehicle.

_____ 2. A type of vehicle jack that uses oil under pressure to lift vehicles.

_____ 3. The maximum safe lifting load for lifting equipment.

_____ 4. A type of vehicle hoist that uses two parts (one on each side of vehicle) and four arms to lift the vehicle.

_____ 5. A type of vehicle hoist that uses a single central platform to lift a vehicle.

_____ 6. A type of hoist onto which a vehicle is driven, that uses two long, narrow platforms to lift the vehicle.

Multiple Choice

Read each item carefully, and then select the best response.

_____ 1. It is good practice for the service advisor to perform a(n) _____ with the customer to point out any existing damage or missing components on the vehicle.
   A. detailed inspection
   B. vehicle walk-around
   C. audit
   D. performance analysis

_____ 2. Battery electrolyte contains acid, which can be cleaned up by neutralizing it with which of the following materials?
   A. An alkaline
   B. Engine oil
   C. Gasoline
   D. Water

_____ 3. There are three main types of mechanisms that provide the lifting action for vehicle jacks including all of the following, *except*:
   A. Hydraulic
   B. Pneumatic
   C. Electric
   D. Mechanical

_____ 4. What type of jack is mounted on four wheels, two of which swivel to provide a steering mechanism?
   A. Bottle jack
   B. Sliding bridge jack
   C. Floor jack
   D. Scissor jack

_____ 5. Often used on farms, what type of jack is designed to lift, winch, clamp, pull, and push?
   A. High-lift jack
   B. Scissor jack
   C. Sliding bridge jack
   D. Bottle jack

_____ 6. What type of jack uses compressed air to either operate a large ram or inflate an expandable air bag to lift the vehicle?
   A. Bottle jack
   B. Air jack
   C. High-lift jack
   D. Floor jack

_____ 7. Which lifting device is useful for raising a vehicle to a height that removes the need for the technician to bend down?
   A. Engine hoist
   B. Jack stand
   C. Farm jack
   D. Vehicle hoist

_____ 8. What type of hoist comes in two configurations—symmetrical and asymmetrical?
   A. Four-post hoist
   B. Two-post hoists
   C. Farm hoist
   D. Single-post hoists

_____ 9. The engine or component to be lifted is attached to the lifting arm of an engine hoist by which of the following?
   A. Sling
   B. Rope
   C. Lifting chain
   D. Either A or C

_____ 10. The size of the vehicle jack you use will be determined by:
   A. The length of the axles
   B. The length of the vehicle
   C. The weight of the vehicle
   D. The type of vehicle

True/False

If you believe the statement to be more true than false, write the letter "T" in the space provided. If you believe the statement to be more false than true, write the letter "F".

_____ **1.** Customers expect their vehicles to be treated with care and respect while in your shop.

_____ **2.** Fender, carpet, seat, and steering wheel covers should be the first thing on and the last thing removed when working on vehicles.

_____ **3.** Never place tools in your back pocket.

_____ **4.** Only the most skilled and experienced drivers available should be allowed to test-drive higher performance vehicles.

_____ **5.** When multiple pieces of lifting equipment are used, the safe working load is determined by the highest rated piece of equipment.

_____ **6.** Vehicle jacks may be used to support the weight of the vehicle during any task that requires you to get underneath any part of the vehicle.

_____ **7.** High-lift jacks are usually fitted in pairs to four-post hoists as an accessory to allow the vehicle to be lifted off the drive-on hoist runways.

_____ **8.** Tall jack stands are used to stabilize a vehicle up on a hoist that is having a heavy component, such as a transaxle, removed or installed.

_____ **9.** Since the vehicle rests on its wheels on the four-post hoist, the wheels cannot be removed, unless the hoist is fitted with sliding bridge jacks.

_____ **10.** Every vehicle hoist in the shop must have a built-in mechanical locking device so that the vehicle hoist can be secured at the chosen height after the vehicle is raised.

Fill in the Blank

Read each item carefully, and then complete the statement by filling in the missing word(s).

1. _____ _____ are a protective layer used to cover the fenders when work is conducted around the engine bay.

2. If spills do occur, be sure to clean them up thoroughly using appropriate methods, which can usually be found in the _____ _____ _____ sheets for each material.

3. Special _____ materials in granular form can be used on some liquid spills such as engine oil.

4. Often _____ _____ will be used by someone directing you to maneuver the vehicle.

5. When using lifting equipment always maintain some _____ _____ as an extra safety margin.

6. Lifting equipment should be periodically _____ and _____ to make sure it is safe.

7. A(n) _____ jack is a portable jack that usually has either a mechanical screw or a hydraulic ram mechanism that rises vertically from the center of the jack as the handle is operated.

8. _____ two-post hoists have arms that are of approximately equal length so that the vehicle is roughly centered lengthwise between the posts.

Labeling
Label the following images with the correct terms.

1. Jacks:

A. _____

B. _____

C. _____

2. Hoists:

A. _____

B. _____

Crossword Puzzle

Use the clues to complete the puzzle.

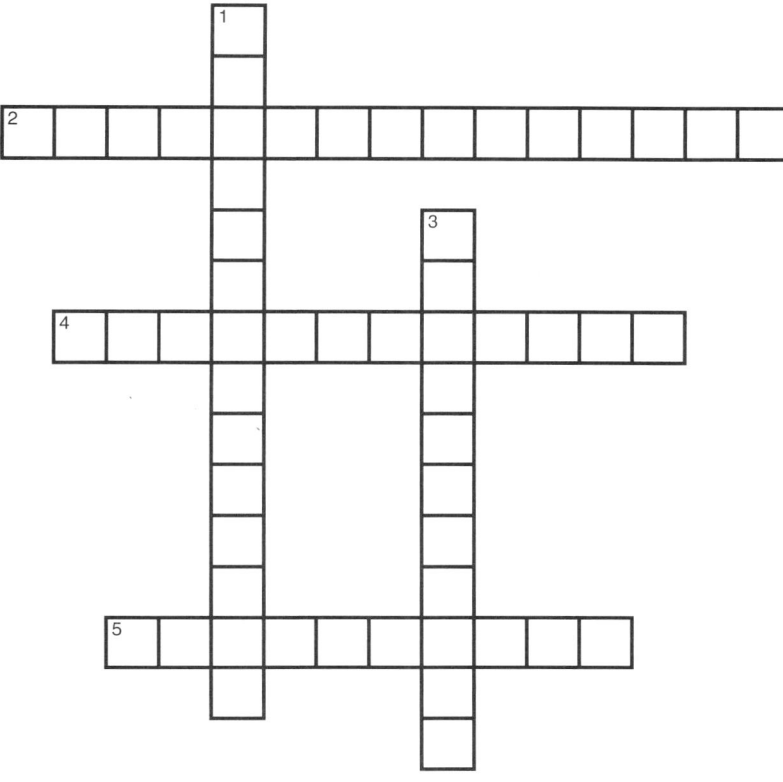

Across

2. A certificate issued when lifting equipment has been checked and deemed safe.

4. A type of vehicle lifting tool designed to lift the entire vehicle.

5. Metal stands with adjustable height to hold a vehicle once it has been jacked up.

Down

1. A type of vehicle jack that uses mechanical leverage to lift a vehicle.

3. A tool for lifting a vehicle.

ASE-Type Questions

Read each item carefully, and then select the best response.

_____ 1. Tech A says that you should always deal with the customer's valuables according to company policy. Tech B says that you should protect a customer's vehicle by washing it when you are finished with it. Who is correct?
  A. Tech A
  B. Tech B
  C. Both A and B
  D. Neither A nor B

_____ 2. Tech A says that it is a good practice to perform a walk-around inspection of the vehicle with the customer. Tech B says that fender covers will protect the fenders from dents when working on the engine. Who is correct?
  A. Tech A
  B. Tech B
  C. Both A and B
  D. Neither A nor B

_____ 3. Tech A says that a vehicle jack can be used to support the vehicle while working under it. Tech B says that a jack stand automatically adjusts to the vehicle's height. Who is correct?
  A. Tech A
  B. Tech B
  C. Both A and B
  D. Neither A nor B

_____ 4. Tech A says that hoists should be inspected and certified periodically. Tech B says that safety locks do not need to be applied before working under the vehicle, unless you will be working for more than 10 minutes. Who is correct?
  A. Tech A
  B. Tech B
  C. Both A and B
  D. Neither A nor B

_____ 5. Tech A says that an engine hoist can lift more weight when the legs and arms are extended. Tech B says that the bolts used to mount an engine to a stand should complete at least six turns. Who is correct?
  A. Tech A
  B. Tech B
  C. Both A and B
  D. Neither A nor B

_____ 6. Tech A says that gasoline vapors are lighter than air, so inspection pits do not have a fire hazard like above ground hoists. Tech B says that you should never lift a fully loaded vehicle off the ground. Who is correct?
  A. Tech A
  B. Tech B
  C. Both A and B
  D. Neither A nor B

_____ 7. Tech A says that you should always ensure that the vehicle has enough ground clearance before driving on a lift. Tech B says that you should center the vehicle on the lift before raising it. Who is correct?
  A. Tech A
  B. Tech B
  C. Both A and B
  D. Neither A nor B

_____ 8. Tech A says that you should always inspect a lifting device for leaks and operation before using it. Tech B says that all mechanical safety locks on a hoist should be in place before getting under the vehicle. Who is correct?
  A. Tech A
  B. Tech B
  C. Both A and B
  D. Neither A nor B

_____ **9.** Tech A says that if you damage a customer's vehicle, insist the damage was there before you started working on it. Tech B says that you should always use seat and steering wheel covers, floor mats, and fender covers whenever a vehicle is being worked on. Who is correct?
- **A.** Tech A
- **B.** Tech B
- **C.** Both A and B
- **D.** Neither A nor B

_____ **10.** Tech A says that a person outside the vehicle should be used to supervise and guide a vehicle while it is being driven in the shop. Tech B says that as long as you are a student in the Auto Tech program you can drive any of the vehicles. Who is correct?
- **A.** Tech A
- **B.** Tech B
- **C.** Both A and B
- **D.** Neither A nor B

Vehicle Maintenance Inspection

Chapter Review

The following activities have been designed to help you refresh your knowledge of this chapter. Your instructor may require you to complete some or all of these activities as a regular part of your training program. You are encouraged to complete any activity that your instructor does not assign as a way to enhance your learning.

Matching

Match the following terms with the correct description or example.

A. Body Control Module (BCM)

B. Coolant

C. CV joint

D. Diesel Exhaust Fluid (DEF)

E. Hydrometer

F. Hygroscopic

G. Pinion shaft

_____ **1.** A property of a substance or liquid that causes it to absorb moisture (water), as a sponge absorbs water.

_____ **2.** A type of universal joint used on the drive axles or half-shafts of a vehicle. Usually refers to front-wheel drive vehicles.

_____ **3.** On a drive axle using a ring-and-pinion gear assembly, the input component that drives the ring gear.

_____ **4.** An onboard computer that controls many vehicle functions including the vehicle interior and exterior lighting, horn, door locks, power seats, and windows.

_____ **5.** The resulting mixture when anti-freeze concentrate is mixed with water.

_____ **6.** A tool that measures the specific gravity of a liquid.

_____ **7.** A mixture of urea and water that is injected into the exhaust system of a late-model diesel-powered vehicle to reduce exhaust nitrogen oxide emissions.

Multiple Choice

Read each item carefully, and then select the best response.

_____ **1.** If the oil in an engine's lubrication system is too high, the oil will be struck by the crankshaft and will:

A. Burn

B. Smoke

C. Coagulate

D. Foam

_____ **2.** Do not remove the _____ when the engine is warm or hot.

A. oil dipstick

B. radiator cap

C. master cylinder cover

D. transmission dipstick

_____ **3.** The anti-freeze protection level can be checked with which of the following?
   A. Hydrometer
   B. Hydroscope
   C. Refractometer
   D. Either A or C

_____ **4.** If an automatic transmission does not have a dipstick, the fluid level is checked using a _____ on the side of the transmission.
   A. master cylinder
   B. reservoir
   C. fill plug
   D. sight glass

_____ **5.** Some late-model diesel-powered vehicles use a fluid called _____, which is injected into the exhaust stream to reduce oxides of Nitrogen during certain driving conditions.
   A. diesel exhaust fluid
   B. oxide reducer
   C. nitrous oxide
   D. octane booster

_____ **6.** What type of drive belt has a flat profile with a number of grooves running lengthwise along the belt?
   A. V-type
   B. Serpentine type
   C. Variable diameter type
   D. Toothed

_____ **7.** The _____ is the most important component driven by the drive belt, and the engine will quickly overheat if the belt breaks or comes off.
   A. alternator
   B. air conditioning compressor
   C. water pump
   D. power steering pump

_____ **8.** What is the shininess on the surface of a belt where it comes in contact with the pulley called?
   A. Shine
   B. Gloss
   C. Soaking
   D. Glazing

_____ **9.** What device on a multiport fuel-injected vehicle is typically located in a rectangular box within the air induction system?
   A. Fuel filter
   B. Air cleaner
   C. Serpentine belt
   D. Carburetor

_____ **10.** What type of fluid is normally green, orange, or yellow in color?
   A. DOT 4
   B. Power steering fluid
   C. Coolant
   D. Automatic transmission fluid

True/False

If you believe the statement to be more true than false, write the letter "T" in the space provided. If you believe the statement to be more false than true, write the letter "F".

_____ 1. Component damage or failure is often caused by a lack of service or low fluid level in the related system.

_____ 2. Always check engine oil with the engine on.

_____ 3. If the brake fluid level gets too low, air can be pulled into the hydraulic system.

_____ 4. Most brake fluid is dark or black.

_____ 5. Some automatic transmissions do not have a dipstick.

_____ 6. Transmission fluid is added through the transmission reservoir.

_____ 7. Many newer vehicles have transmissions that are considered sealed and lubricated for the life of the vehicle.

_____ 8. There are two types of drive belts: the V-type and the serpentine type.

_____ 9. The engine should be hot when inspecting radiator hoses.

_____ 10. The warning lamps perform a self-check each time the ignition is switched on or the engine is cranked.

Fill in the Blank

Read each item carefully, and then complete the statement by filling in the missing word(s).

1. Nearly all of the _____ in a vehicle, except some used in automatic transmissions/transaxles get old and wear out, requiring replacement.

2. The level of _____ in an engine's lubrication system is critical to the engine's operation.

3. Engine _____ must be controlled to prevent overheating and to maintain proper exhaust emission levels.

4. A hydraulic braking system depends on a special fluid called _____ _____.

5. When checking power steering fluid levels the engine should be _____ and the fluid hot.

6. Driving in dusty or wet conditions may require that the _____ _____ _____ be checked more often.

7. Engine _____ _____ are used to operate the various accessories on the engine, such as the water pump, power steering pump, air conditioner compressor, and alternator.

8. _____ that exceed a certain number per inch in a belt indicate that the belt may soon fail and should be replaced.

9. The _____ _____ should flex at the hinge and be held firmly against the windshield by the wiper arm spring.

10. The _____ _____ acts as a lever to increase the force applied to the brake assemblies by the driver.

Labeling

Label the following diagrams with the correct terms.

Power steering fluid reservoir:

A. _____   B. _____

ASE-Type Questions

Read each item carefully, and then select the best response.

_____ **1.** Tech A says that during an exterior inspection of a vehicle you should open and close doors to check that they are operating correctly. Tech B says you should never pull on bumpers or fenders. Who is correct?
   A. Tech A
   B. Tech B
   C. Both A and B
   D. Neither A nor B

_____ **2.** Tech A says wiper blade condition cannot cause a vehicle to fail a safety inspection in some states. Tech B says check wiper blades for wear and tear. Who is correct?
   A. Tech A
   B. Tech B
   C. Both A and B
   D. Neither A nor B

_____ **3.** Tech A says that a low oil level is bad for the engine but that it is OK for the level to be too high. Tech B says that overfilling the engine oil is bad for the engine. Who is correct?
   A. Tech A
   B. Tech B
   C. Both A and B
   D. Neither A nor B

_____ 4. Tech A says that improper handling of a windshield wiper can lead to a broken windshield. Tech B says you should place a fender cover on the windshield to prevent damage while working on windshield wipers. Who is correct?
   A. Tech A
   B. Tech B
   C. Both A and B
   D. Neither A nor B

_____ 5. While servicing an automatic transaxle–equipped vehicle, the transaxle dipstick cannot be found. Tech A says that some of these vehicles do not have a transaxle dipstick. Tech B says that the dipstick may have fallen off, since all vehicles have a transaxle dipstick. Who is correct?
   A. Tech A
   B. Tech B
   C. Both A and B
   D. Neither A nor B

_____ 6. Tech A says to test new wiper blades against a dry windshield to ensure they seat properly. Tech B says to wet the windshield and operate the wipers to check their performance. Who is correct?
   A. Tech A
   B. Tech B
   C. Both Tech A and Tech B
   D. Neither Tech A nor Tech B

_____ 7. An engine serpentine belt has broken and come off of the pulleys. Tech A says that it is OK to drive the vehicle without the belt. Tech B says that the belt only runs the charging system, so it is OK to drive a short distance without the belt. Who is correct?
   A. Tech A
   B. Tech B
   C. Both A and B
   D. Neither A nor B

_____ 8. Tech A says to clean a windshield first, and then inspect the windshield for chips, scratches, or etching. Tech B says to always clean a windshield first and then inspect the windshield for cracks or signs of delamination. Who is correct?
   A. Tech A
   B. Tech B
   C. Both A and B
   D. Neither A nor B

_____ 9. Tech A says that unlocking doors on most vehicles turns on the interior lights. Tech B says on some vehicles the interior light will not turn on until a door is actually opened. Who is correct?
   A. Tech A
   B. Tech B
   C. Both A and B
   D. Neither A nor B

_____ 10. Tech A says you should test the rear lights with the help of an assistant. Tech B says some shops have a mirror mounted in the service bay so the technician can check rear lights. Who is correct?
   A. Tech A
   B. Tech B
   C. Both A and B
   D. Neither A nor B

Communication

Chapter Review

The following activities have been designed to help you refresh your knowledge of this chapter. Your instructor may require you to complete some or all of these activities as a regular part of your training program. You are encouraged to complete any activity that your instructor does not assign as a way to enhance your learning.

Multiple Choice

Read each item carefully, and then select the best response.

_____ **1.** The first step in good communication is:
- **A.** Nonverbal communication
- **B.** Asking good questions
- **C.** Providing feedback
- **D.** Active listening

_____ **2.** Attempting to see a situation from someone else's point of view is known as which of the following?
- **A.** Sympathy
- **B.** Empathy
- **C.** Nonverbal communication
- **D.** Feedback

_____ **3.** Phrases such as "I see" or "Tell me more" that indicate you are paying attention are examples of a(n)_____?
- **A.** validating statement
- **B.** supporting statement
- **C.** closed question
- **D.** open question

_____ **4.** Statements like "Go on" and "Give me an example" that let the speaker know you would like more detail because you are genuinely interested in finding a solution are examples of a(n)_____?
- **A.** validating statement
- **B.** supporting statement
- **C.** closed question
- **D.** open question

_____ **5.** What type of question usually begins with the words when or where?
- **A.** Open questions
- **B.** Yes/no questions
- **C.** Closed questions
- **D.** Good questions

_____ **6.** What aspect of our physical appearance creates the first impression made in any encounter?
- **A.** Clothes
- **B.** Hairstyle
- **C.** Demeanor
- **D.** All of the above

_____ **7.** Looking through the table of contents, introduction, conclusion, headings, and index until we find what we are looking for is the quickest way to use what type of reading?
  A. Absorbing
  B. Comprehending
  C. Selective
  D. Quick

_____ **8.** Which of the following is like conducting an investigation?
  A. Reading
  B. Researching
  C. Comprehending
  D. Skimming

_____ **9.** What online reference organization consists of over 75,000 active members, with more than 1.7 million years of combined experience, who share their knowledge with each other on over a dozen forums?
  A. International Automotive Technicians Network
  B. National Association for Stock Car Auto Racing
  C. Occupational Safety and Health Administration
  D. National Automotive Parts Association

_____ **10.** Which of the following should you do when you come across any defective equipment?
  A. Tag the defective item
  B. Complete a defective equipment report
  C. Notify your supervisor
  D. All of the above

True/False

If you believe the statement to be more true than false, write the letter "T" in the space provided. If you believe the statement to be more false than true, write the letter "F".

_____ **1.** Learning and applying good communication skills will save you time and help you avoid or get through tricky situations.

_____ **2.** Most people tend to believe the actual words expressed over the nonverbal message you are sending.

_____ **3.** Your body position, eye contact, and facial expression can all set the direction of a conversation.

_____ **4.** Speaking is often referred to as a science.

_____ **5.** Closed questions allow individuals to answer with a simple yes or no.

_____ **6.** Nonverbal communication is not possible during a phone conversation.

_____ **7.** Being part of a team allows us to complement each other's strengths and weaknesses.

_____ **8.** If you have to take customers into the shop, always escort them, and be sure to keep them safe.

_____ **9.** Routinely texting your friends or taking personal calls during work hours is stealing from your employer.

_____ **10.** The CSI rating is reported each month and used to evaluate individual technicians and the entire service facility.

_____ **11.** The person who wants the job done right will probably be very keen to have the repairs finished on schedule.

_____ **12.** The faster you try to read, the more you are likely to be able to concentrate on the meaning.

_____ **13.** A repair order can become a legal document that will be used by the court to determine if the shop has any liability in a lawsuit situation.

_____ **14.** Safety inspection forms are a way of keeping up on routine maintenance tasks such as changing the oil or replacing the belt of an air compressor.

_____ **15.** If a problem is noticed with a piece of equipment, the lockout/tagout procedure should be followed.

Fill in the Blank

Read each item carefully, and then complete the statement by filling in the missing word(s).

1. Mental _____ are thoughts and feelings that interfere with our listening, such as our own assumptions, emotions, and prejudices.

2. As a person is speaking to you, you need to provide _____ feedback, which indicates to the speaker that you are engaged in what he or she is saying.

3. _____ feedback includes very simple spoken signals that can enhance the conversation and let the person know you comprehend.

4. Before speaking, take a moment to consider that your _____ of _____ reveals a lot about your feelings and adds significant meaning to your message.

5. A(n) _____ question encourages people to speak freely so we can gather facts, insights, and opinions from them.

6. When taking a _____ _____ for someone else, make sure you have the caller's name and organization, contact details, the date and time of the call, and a summary of the caller's message.

7. _____ should contain information about who, what, when, where, and why, and direction on how a task should be completed.

8. Good _____ skills involve setting a clear vision of the goals, empowering each team member to contribute his or her best efforts, and recognizing each team member's strengths and weaknesses.

9. Our _____ is the image we present of ourselves to the public.

10. _____ means showing up on time (typically 5–10 minutes early to get ready to start work at the appointed time).

11. _____ reading is reading only the parts we need to know. This method is useful when looking for a particular piece of information.

12. Before you spend too much time researching information, it is important that you _____ the problem.

13. When completing a repair order these elements constitute what are called the three Cs: _____, _____, and _____.

14. If someone is injured in the workplace a(n) _____ _____ should be completed by those involved, both the victim and witnesses, if possible.

15. When performing a(n) _____, you need to check that all major components and systems are operational, secured, and safe in accordance with the vehicle manufacturer's recommendations.

ASE-Type Questions

Read each item carefully, and then select the best response.

_____ 1. Tech A says that repair orders are legal documents so they need to be filled out accurately and carefully. Tech B says that ensuring the repair order is well written, clear, and concise promotes a professional reputation. Who is correct?
   A. Tech A
   B. Tech B
   C. Both A and B
   D. Neither A nor B

_____ 2. Tech A says that you shouldn't waste time inspecting a vehicle in for a repair unless the customer requests it. Tech B says that performing an inspection in addition to a repair can lead to discovery of additional concerns. Who is correct?
   A. Tech A
   B. Tech B
   C. Both A and B
   D. Neither A nor B

_____ **3.** Tech A says that the proper way to listen to a customer is to maintain eye contact with the customer in between taking notes. Tech B says that eye contact distracts you from taking good notes. Who is correct?
   A. Tech A
   B. Tech B
   C. Both A and B
   D. Neither A nor B

_____ **4.** Tech A says it is best to quietly listen to the customer and refrain from asking questions as much as possible. Tech B says it is constructive to ask questions throughout the conversation to obtain more details. Who is correct?
   A. Tech A
   B. Tech B
   C. Both A and B
   D. Neither A nor B

_____ **5.** Tech A says that maintaining an appearance of neatness is important as it conveys to the customer the idea of careful, professional technicians. Tech B says that a dirty and cluttered shop indicates that the shop gets a lot of quality work done. Who is correct?
   A. Tech A
   B. Tech B
   C. Both A and B
   D. Neither A nor B

_____ **6.** Tech A says researching the service information is a waste of time. Tech B says that researching the service information saves time. Who is correct?
   A. Tech A
   B. Tech B
   C. Both A and B
   D. Neither A nor B

_____ **7.** Tech A says that an example of an open question is: "What are the conditions like when your A/C is not working?" Tech B says that an example of an open question is: "Does your A/C work at all?" Who is correct?
   A. Tech A
   B. Tech B
   C. Both A and B
   D. Neither A nor B

_____ **8.** Tech A says that one customer who has a bad experience stemming from miscommunication and a misdiagnosed repair due to a technician's failure to listen has more of an effect on the repair shop than several good and happy customers. Tech B says that it is only one customer, and since the happy ones paid their bills, all is well. Who is correct?
   A. Tech A
   B. Tech B
   C. Both A and B
   D. Neither A nor B

_____ **9.** Tech A says that it is primarily the job of the service writer to be customer-oriented and learn listening skills. Tech B says that customer service is the responsibility of all service employees to achieve the goal of customer satisfaction. Who is correct?
   A. Tech A
   B. Tech B
   C. Both A and B
   D. Neither A nor B

_____ **10.** Tech A says that the 3 Cs are the "customer, complaint, and concern." Tech B says that the 3 Cs are "concern, cause, and correction." Who is correct?
   A. Tech A
   B. Tech B
   C. Both A and B
   D. Neither A nor B

Motive Power Types—Spark-Ignition (SI) Engines

Chapter Review

The following activities have been designed to help you refresh your knowledge of this chapter. Your instructor may require you to complete some or all of these activities as a regular part of your training program. You are encouraged to complete any activity that your instructor does not assign as a way to enhance your learning.

Matching

Match the following terms with the correct description or example.

A. Base circle
B. Bottom dead center (BDC)
C. Cam
D. Camshaft
E. Column inertia
F. Cylinder bore
G. Duration
H. Exhaust stroke
I. Horsepower

J. Ignition
K. Lobe
L. Piston clearance
M. Piston skirt
N. Power
O. Spark ignition (SI) engine
P. Torque
Q. Valve margin
R. Work

_____ 1. The hole in the engine block that the piston fits into.

_____ 2. An engine that relies on an electrical spark to ignite the air and fuel mixture.

_____ 3. The stroke of piston during which the exhaust valve is open and the piston is moving from bottom dead center to top dead center to push exhaust gas out of the cylinder.

_____ 4. The clearance between the piston and the cylinder wall that allows for lubricating oil to reduce friction.

_____ 5. The flat surface on the outer edge of the valve head between the valve head and the valve face.

_____ 6. The principle that as a column of air flows, it creates inertia, which keeps air flowing until its inertia energy is spent; sometimes referred to as a "ram effect" when using tuned intake or exhaust systems.

_____ 7. The amount of twisting force applied in a turning application, usually measured in foot-pounds.

_____ 8. The egg-shaped lobe machined to a shaft used to cause opening and closing of the valves of a four-stroke cycle engine.

_____ 9. The amount of time the valve stays open, given in degrees of rotation of the crankshaft.

_____ 10. The position of the piston at the end of its stroke when it is closest to the crankshaft.

_____ 11. The part of the engine that activates the valve train by using lobes riding against lifters.

_____ 12. The lighting of the fuel and air mixture in the combustion chamber.

_____ 13. The result of force creating movement.

_____ 14. The area below the ring groove area of the piston that prevents the piston from cocking and becoming jammed in the cylinder bore.

_____ 15. The rounded bottom part of the camshaft (off the lobe) where the valves remain closed or at rest.

_____ 16. The rate or speed at which work is done.

_____ 17. An amount of work performed in a given time.

_____ 18. The raised portion on a camshaft; used to lift the lifter and open the valve.

Multiple Choice

Read each item carefully, and then select the best response.

_____ 1. The steam engine and the Stirling engine are examples of what kind of engine?
   A. Internal combustion engine
   B. External combustion engine
   C. Rotary engine
   D. Reciprocating piston engine

_____ 2. Which of the following types of fuel contains latent heat energy?
   A. Diesel
   B. Ethanol
   C. Hydrogen
   D. All of the above

_____ 3. The effort to produce a push or pull action is referred to as _____.
   A. power
   B. work
   C. force
   D. torque

_____ 4. The unit of measurement for torque in the metric system is _____.
   A. ft-lb
   B. Newton meters
   C. horsepower
   D. rpms

_____ 5. How many radians are there in the circumference of a circle?
   A. 1
   B. 3.14
   C. 6.28
   D. Depends on the radius of the circle

_____ 6. The _____ occurs as extreme force moves the piston from TDC to BDC with both valves remaining closed.
   A. power stroke
   B. compression stroke
   C. intake stroke
   D. exhaust stroke

_____ 7. The distance the piston travels from TDC to BDC, or from BDC to TDC, is called the _____.
   A. throw
   B. piston stroke
   C. displacement
   D. compression ratio

_____ 8. The displacement of an engine can be altered by changing the _____.
   A. cylinder bore
   B. piston stroke
   C. number of cylinders
   D. all of the above

_____ 9. The Miller cycle engine and the Atkinson cycle engine are both variations on the traditional _____ engine.
   A. two-stroke SI
   B. four-stroke SI
   C. external combustion engine
   D. rotary engine

_____ **10.** The _____ are the areas between the ring grooves that support the rings as the piston moves.
  A. offsets
  B. journals
  C. ring lands
  D. skirts

_____ **11.** The piston _____ prevents the piston from rocking and jamming in the cylinder bore.
  A. skirt
  B. pin boss
  C. connecting rod
  D. head

_____ **12.** What type of engines use pushrods to transfer the camshaft's lifting motion to the valves by way of rocker arms on top of the cylinder head?
  A. Flathead engines
  B. Cam-in-block engines
  C. Overhead cam
  D. Dual overhead cam (DOHC) engines

_____ **13.** Which part rides on the camshaft lobes to actuate the pushrods, rocker arms, and valves?
  A. Tappets
  B. B Ramps
  C. Lifters
  D. Either A or C

_____ **14.** The number of degrees between the centerline of the intake lobe and the centerline of the exhaust lobe is called _____.
  A. cam lobe centerline
  B. cam lobe separation
  C. cam lobe duration
  D. cam lobe margin

_____ **15.** A machined surface on the back of the valve head is known as the valve _____.
  A. face
  B. margin
  C. keeper
  D. stem

_____ **16.** The valve train operates off of the camshaft, and the part that rides against the cam lobe is the _____.
  A. valve stem
  B. valve lifter
  C. rocker arm
  D. fulcrum

_____ **17.** The _____ engine has enough clearance between the pistons and the valves so in the event the timing belt breaks any valve that is hanging all the way open will not contact the piston, thus preventing engine damage.
  A. interference
  B. rotary
  C. freewheeling
  D. stirling

_____ **18.** The _____ engine is also called the Wankle engine because it was improved upon by Felix Wankle for automotive use in the 1940s.
  A. two-stroke
  B. four-stroke
  C. interference
  D. rotary

True/False

If you believe the statement to be more true than false, write the letter "T" in the space provided. If you believe the statement to be more false than true, write the letter "F".

_____ **1.** The internal combustion engine has almost completely replaced the external combustion engine and has been around for well over a century.

_____ **2.** Compression-ignition engines do not use spark plugs.

_____ **3.** Pressure and volume are inversely related; as one rises, the other falls.

_____ **4.** Force is measured in ft-lb per second or ft-lb per minute.

_____ **5.** Movement must occur to produce torque.

_____ **6.** A turbocharger or supercharger will increase an engine's volumetric efficiency well above 100%.

_____ **7.** When the piston in the cylinder is at a position closest to the crankshaft, it is said to be at top dead center.

_____ **8.** The power stroke starts with the exhaust valve closed, the intake valve(s) opening, and the piston moving from TDC to BDC.

_____ **9.** The block deck is the machined surface of the block farthest from the crankshaft.

_____ **10.** Increasing the diameter of the bore or increasing the length of the stroke will produce a larger piston displacement.

_____ **11.** The time that both the intake and exhaust valves are open is called valve overlap.

_____ **12.** A long block replacement includes the engine block from below the head gasket to above the oil pan.

_____ **13.** The piston body has a piston pinhole machined through a reinforced area called the piston pin boss.

_____ **14.** On many engines, the oil pan houses the oil pump, which is the heart of the engine in that it supplies critical lubrication and cooling for the internal moving parts of the engine.

_____ **15.** Intake and exhaust valves may all be actuated by a single camshaft, or there may be two camshafts per head, called dual overhead cam engines.

_____ **16.** The cam lobe ramp is where the rise of the cam lobe starts from the base circle to the top of the lobe.

_____ **17.** If valve clearance is too large, the valve can be held open longer than it should be.

_____ **18.** The rotary engine is not an internal combustion engine.

Fill in the Blank

Read each item carefully, and then complete the statement by filling in the missing word(s).

1. The branch of physical science that deals with heat and its relation to other forms of energy such as mechanical energy is generally defined as _____.

2. The gasoline _____ engine uses a crankshaft to convert the reciprocating movement of the pistons in their cylinder bores into rotary motion at the crankshaft.

3. Heating a gas in a sealed container will _____ the pressure in the container.

4. One _____ equals 33,000 ft-lb per minute.

5. A(n) _____ describes how many radius distances there are in the circumference of a circle.

6. When the piston in a cylinder is at the position farthest away from the crankshaft, it is at _____ _____ _____.

7. Internal combustion engines are designated by the amount of space their pistons displace as they move from TDC to BDC, which is called _____ _____.

8. The process of using a column of moving air to create a low-pressure area behind it to assist in removing any remaining burned gases from the combustion chamber and replacing these gases with a new charge is called _____.

9. The _____ _____ is the single largest part of the engine.

10. A(n) _____ is a weighted assembly that stores kinetic energy from each power stroke and helps keep the crankshaft turning through nonpower strokes.

11. The _____ _____ connects the piston to the crankshaft and transfers piston movement and combustion pressure to the crankshaft rod journals.

12. The upper two piston rings are compression rings, which prevent combustion pressure, called _____ _____, from leaking past the pistons into the crankcase.

13. A(n) _____ is a combination of materials that has properties that are different from the original materials.

14. Up until the 1950s, many engines had their valves installed in the engine block. Such engines are called _____ engines.

15. Coil _____ occurs when the coils of the spring touch each other.

16. The opposite end of the valve stem has grooves machined into its end to receive locking pieces, sometimes referred to as valve _____, that retain a valve spring retainer and spring on the valve.

17. In a _____-_____ engine, the inlet and exhaust ports are opened and closed by the movement of the piston.

18. A(n) _____ _____ is the circular movement around the perimeter of another circle.

Labeling

Label the following diagrams with the correct terms.

1. External combustion engines:

Steam engine

A. _____

B. _____

C. _____

D. _____

E. _____

F. _____

Stirling engine

A. _____

B. _____

C. _____

D. _____

E. _____

F. _____

2. Temperature changes pressure:

A: As temperature
   increases, the
   pressure increases

B: As temperature
   decreases, the
   pressure decreases

A. _____

B. _____

C. _____

D. _____

3. Volume affects pressure:

A: As the volume
   decreases, the
   pressure increases

B: As the volume
   increases, the
   pressure decreases

A. _____

B. _____

C. _____

D. _____

E. _____

4. The basic four-stroke cycle:

A. _____

B. _____

C. _____

D. _____

E. _____

5. Engine block:

A. _____

B. _____

C. _____

D. _____

E. _____

F. _____

G. _____

H. _____

I. _____

J. _____

K. _____

6. Basic parts of the crankshaft:

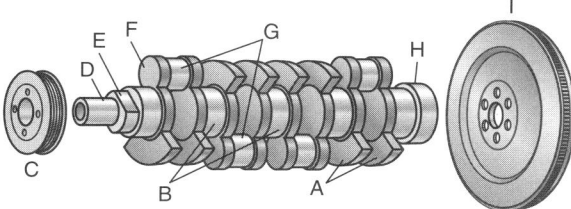

A. _____

B. _____

C. _____

D. _____

E. _____

F. _____

G. _____

H. _____

I. _____

7. Piston and piston rings:

A. _____

B. _____

C. _____

D. _____

E. _____

F. _____

G. _____

H. _____

I. _____

J. _____

K. _____

L. _____

M. _____

N. _____

8. Camshaft lobe:

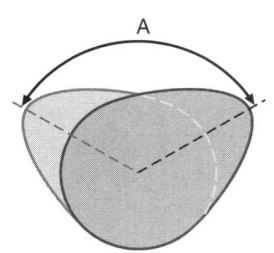

A. _____

B. _____

C. _____

D. _____

E. _____

F. _____

G. _____

9. Two-stroke reed valve operation:

A. _____

B. _____

C. _____

D. _____

E. _____

10. Rotary engine:

1. _____

2. _____

3. _____

4. _____

5. _____

A. _____

B. _____

C. _____

D. _____

E. _____

F. _____

G. _____

H. _____

Crossword Puzzle

Use the clues to complete the puzzle.

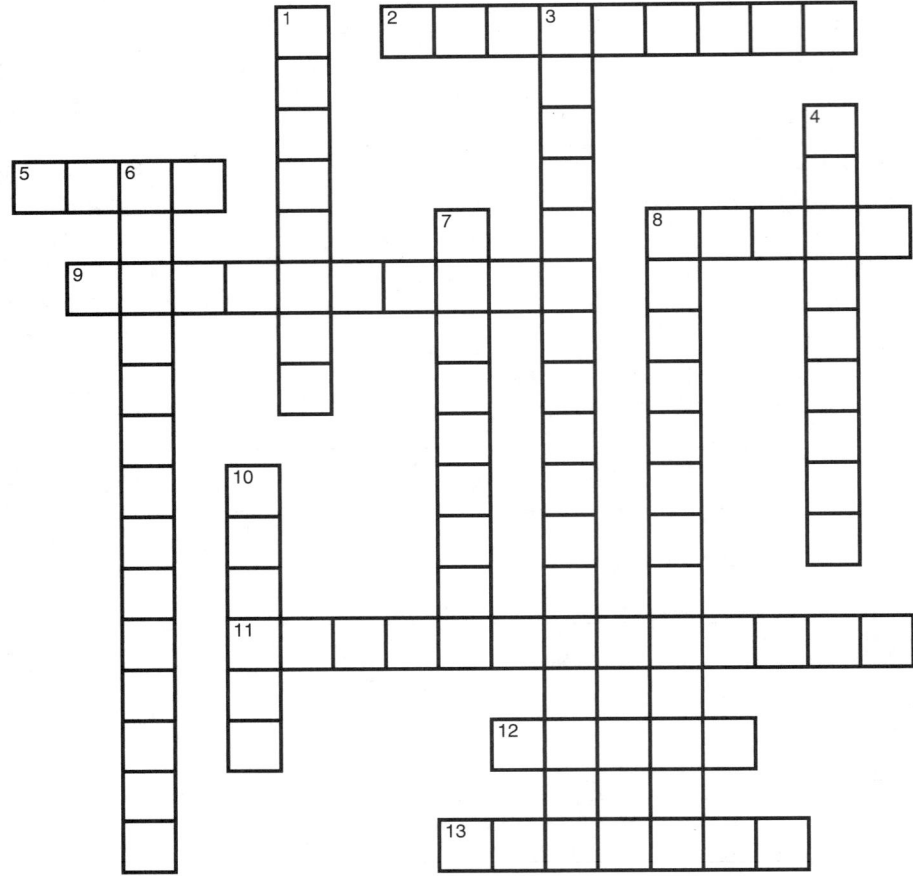

Across

2. The "top" of the engine block and cylinder bore where the cylinder head is bolted on.
5. The amount the valve will open. The more the valve lifts off its seat, the more air can get into and out of the engine.
8. The mixture of materials to make a substance that has properties different from the original materials.
9. An insert in the cylinder head through which the valve stem passes and moves.
11. A loss of engine efficiency caused by internal friction, inefficient breathing, etc.
12. The effort to produce a push or pull action.
13. Locking devices that keep the valve retained by the valve spring seat.

Down

1. The heavy, circular flat plate that keeps the engine rotating when power is not produced, such as on the exhaust, intake, and compression strokes.
3. The stroke of the piston during which air and fuel is being compressed into a small area prior to ignition.
4. The result of combustion gases leaking past the compression rings and getting into the crankcase.
6. An L-head engine with valves in the block.
7. The metal between the ring grooves that supports the rings as the piston moves.
8. An engine cycle that uses a longer effective exhaust stroke than intake stroke to reduce exhaust emissions. This type of engine is widely used in hybrid-electric vehicles.
10. Another name for a valve lifter. It may be flat or have rollers to ride on the cam lobes.

ASE-Type Questions

Read each item carefully, and then select the best response.

_____ 1. Tech A says that engines using compression ignition control timing by regulating when fuel is injected into the cylinder. Tech B says that engines using spark ignition control timing by regulating when fuel is injected into the cylinder. Who is correct?
   A. Tech A
   B. Tech B
   C. Both A and B
   D. Neither A nor B

_____ 2. Tech B says that horsepower is a measurement of the amount of work being performed. Tech B says that horsepower can be calculated by multiplying torque by rpm and dividing by 5252. Who is correct?
   A. Tech A
   B. Tech B
   C. Both A and B
   D. Neither A nor B

_____ 3. Tech B says that in a four-stroke engine, the piston is at TDC four times to complete the cycle. Tech B says that the air/fuel mixture is ignited once every two strokes. Who is correct?
   A. Tech A
   B. Tech B
   C. Both A and B
   D. Neither A nor B

_____ 4. Tech A says that spark ignition typically occurs before TDC. Tech B says that spark ignition typically occurs after TDC. Who is correct?
   A. Tech A
   B. Tech B
   C. Both A and B
   D. Neither A nor B

_____ 5. Tech A says that valve overlap occurs between the exhaust stroke and the intake stroke. Tech B says that valve overlap occurs to assist in scavenging the cylinder. Who is correct?
   A. Tech A
   B. Tech B
   C. Both A and B
   D. Neither A nor B

_____ 6. Tech A says that compression ratio is the comparison of the volume above the piston at BDC to the volume above the piston at TDC. Tech B says that scavenging of the exhaust gases occurs once the exhaust valve closes. Who is correct?
   A. Tech A
   B. Tech B
   C. Both A and B
   D. Neither A nor B

_____ 7. Tech A says that the weight of the flywheel smooths out the engine's power pulses. Tech B says that the flex plate and torque converter perform the same function as the flywheel. Who is correct?
   A. Tech A
   B. Tech B
   C. Both A and B
   D. Neither A nor B

_____ 8. Tech A says that an interference engine is designed so that the pistons can hit the valves if the timing belt breaks. Tech B says that the shape of the cam lobe determines how long and far the valves are held open. Who is correct?
   A. Tech A
   B. Tech B
   C. Both A and B
   D. Neither A nor B

_____ **9.** Tech A says that blowby gases occur when compression and combustion gases leak past the piston rings. Tech B says that a rotary engine uses reed valves to control air flow into the cylinder on intake. Who is correct?

    A. Tech A

    B. Tech B

    C. Both A and B

    D. Neither A nor B

_____ **10.** Tech A says that the principle of thermal expansion is what pushes the piston down the cylinder on the power stroke. Tech B says that the piston is pulled down the cylinder on the intake stroke. Who is correct?

    A. Tech A

    B. Tech B

    C. Both A and B

    D. Neither A nor B

Engine Mechanical Testing

Chapter Review

The following activities have been designed to help you refresh your knowledge of this chapter. Your instructor may require you to complete some or all of these activities as a regular part of your training program. You are encouraged to complete any activity that your instructor does not assign as a way to enhance your learning.

Matching

Match the following terms with the correct description or example.

A. Compression tester
B. Cylinder leakage tester
C. Data link connector (DLC)

D. Pressure transducer
E. Vacuum gauge

_____ 1. The connector through which the scan tool communicates to the vehicle's computers; it will display the readings from the various sensors and can retrieve trouble codes, freeze-frame data, and system monitor data.

_____ 2. A device used to measure the amount of pressure a cylinder can generate.

_____ 3. A device used to measure the amount of vacuum an engine can generate during various operating conditions.

_____ 4. A device that pumps air into the cylinder and measures the percentage of air that is leaking out of the cylinder.

_____ 5. A device used to measure engine vacuum and display it graphically on a lab scope.

Multiple Choice

Read each item carefully, and then select the best response.

_____ 1. What tool, available in both standard and electronic versions is used by technicians to listen for unusual noises in a vehicle?
A. Microphone
B. Stethoscope
C. Sound transducer
D. Amplifier

_____ 2. When using fluorescent dye to help locate the leak, the dye can only be seen under what kind of light?
A. Fluorescent
B. Infrared
C. Ultraviolet
D. Incandescent

_____ 3. Many newer vehicles are programmed with a _____ capability that effectively shuts off the fuel injectors as long as the throttle is held to the floor before the ignition key is turned to the run or crank position.
   A. time delay
   B. cold start
   C. crank mode
   D. clear flood

_____ 4. A low and steady reading gauge reading during a vacuum test indicates?
   A. Everything is good
   B. Possible late valve or ignition timing
   C. Possible worn rings
   D. Burned or leaking valve

_____ 5. The typical vacuum reading for a properly running engine at idle is a steady _____ of vacuum.
   A. 10" to 12"
   B. 17" to 21"
   C. 4" to 8"
   D. 20" to 23"

_____ 6. Using a pressure transducer and _____ is a similar process to using a vacuum gauge, except it is much more accurate and allows you to look at the vacuum graphically.
   A. lab scope
   B. stethoscope
   C. amp probe
   D. monitor

_____ 7. Which test identifies which cylinder(s) are not operating properly and gives a general indication of each cylinder's overall health?
   A. Vacuum gauge test
   B. Cylinder leakage test
   C. Pressure transducer test
   D. Power balance test

_____ 8. In which test is a high-pressure hose hand threaded into the spark plug hole of the cylinder to be tested and then connected to the compression gauge?
   A. Power balance test
   B. Vacuum gauge test
   C. Cranking or running compression test
   D. Cylinder leakage test

_____ 9. What test checks an engine's ability to move air into and out of the cylinder?
   A. Running compression test
   B. Vacuum gauge test
   C. Power balance test
   D. Running compression test

_____ 10. Black exhaust is an indication of?
   A. Coolant leak
   B. Excessive rich fuel mixture
   C. Burning engine oil
   D. Blown head gasket

True/False

If you believe the statement to be more true than false, write the letter "T" in the space provided. If you believe the statement to be more false than true, write the letter "F".

_____ 1. Mechanical testing starts with a good visual inspection.

_____ 2. A mechanic could diagnose a burned valve with just a piece of bubble gum and its wrapper.

_____ 3. The smell of a leaking fluid can give you a clue as to its source.

_____ 4. Seals leak only when the engine is running.

_____ 5. You can disable the ignition system so that it will crank but not start.

_____ 6. A sharp oscillation back and forth in the needle or a dip in the vacuum gauge reading could relate to a problem such as a restricted exhaust system.

_____ 7. Using a vacuum gauge is much more accurate than using a pressure transducer and lab scope.

_____ 8. Disabling a cylinder that is not operating correctly will not produce much, if any rpm drop.

_____ 9. A cranking compression test is performed when indications show a misfiring or dead cylinder that is not caused by an ignition or fuel problem.

_____ 10. A rod bearing makes an evenly spaced single knock while a main bearing generally makes a double knock.

Fill in the Blank

Read each item carefully, and then complete the statement by filling in the missing word(s).

1. Engine mechanical testing uses a series of tests to assess the mechanical _____ of the engine.

2. Sometimes adding a special fluorescent _____ to the fluid will help locate a leak.

3. If _____ is not the similar across all cylinders, a tune-up will not fix the problem.

4. A(n) _____ _____ can be used so that the vacuum gauge can be connected and, at the same time, vacuum can still be supplied to the existing component.

5. A vacuum _____ reading shows the difference between outside atmospheric pressure and the amount of manifold pressure in the engine.

6. During an engine vacuum test, all vacuum gauges are calibrated to, and all instructions and readings are referenced to _____ _____.

7. The purpose of the _____ _____ test is to see whether the cylinders are creating equal amounts of power and, if not, to isolate the problem to a particular cylinder or cylinders.

8. When a low-compression cylinder is found, you should put a couple squirts of oil into the spark plug hole, crank the engine a few turns, and recheck the compression. if the compression rises significantly, the problem is typically worn _____ _____.

9. The _____ _____ test is performed by leaving all of the spark plugs in the engine except for the one in the cylinder that you are testing.

10. The _____ _____ test is performed on a cylinder with low compression to determine the severity of the compression leak and where the leak is located.

Labeling

Label the following images with the correct terms.

Testing tools:

A. _____

B. _____

C. _____

D. _____

ASE-Type Questions

Read each item carefully, and then select the best response.

_____ **1.** Tech A says that a cranking sound diagnosis can be used to diagnose problems in the ignition system. Tech B says that a cranking sound diagnosis can indicate differences in compression. Who is correct?
  A. Tech A
  B. Tech B
  C. Both A and B
  D. Neither A nor B

_____ **2.** Tech A says that a power balance test is a good way to narrow a misfire down to a particular cylinder or cylinders. Tech B says that a cylinder power balance test measures the volumetric efficiency of the cylinder being tested. Who is correct?
  A. Tech A
  B. Tech B
  C. Both A and B
  D. Neither A nor B

_____ **3.** Tech A says that a cranking compression wet test can indicate if the cylinder has worn piston rings. Tech B says that the throttle should be held wide open during a cranking compression check. Who is correct?
   A. Tech A
   B. Tech B
   C. Both A and B
   D. Neither A nor B

_____ **4.** Tech A says that low compression on a single cylinder will cause an engine not to start. Tech B says that low compression on a single cylinder will affect the engine's cranking sound. Who is correct?
   A. Tech A
   B. Tech B
   C. Both A and B
   D. Neither A nor B

_____ **5.** Tech A says that a cylinder leakage test is performed on a cylinder with low compression to determine the severity of the leak and where it is located. Tech B says that manufacturers will consider up to 50% cylinder leakage past the piston rings acceptable. Who is correct?
   A. Tech A
   B. Tech B
   C. Both A and B
   D. Neither A nor B

_____ **6.** Tech A says that a scan tool connected to the data link connector (DLC) will perform a cylinder power balance test and report cylinder pressures. Tech B says that a scan tool will perform a cylinder power balance test and report whether the rings or valves have failed. Who is correct?
   A. Tech A
   B. Tech B
   C. Both A and B
   D. Neither A nor B

_____ **7.** Tech A says that researching related service information for a vehicle repair will assist the technician. Tech B says that research should be done only when the technician needs direction. Who is correct?
   A. Tech A
   B. Tech B
   C. Both A and B
   D. Neither A nor B

_____ **8.** Tech A says that a vacuum gauge needle that dips 4–8 inches rhythmically can indicate a burned valve. Tech B says that a stethoscope can be used to determine the source of unusual engine noises. Who is correct?
   A. Tech A
   B. Tech B
   C. Both A and B
   D. Neither A nor B

_____ **9.** Tech A says that a vacuum test can determine exhaust restriction. Tech B says that when performing a cylinder power balance test, results should be 5% or less. Who is correct?
   A. Tech A
   B. Tech B
   C. Both A and B
   D. Neither A nor B

_____ **10.** Tech A says that a bad cam lobe or broken valve spring will show up during a running compression test. Tech B says that when performing a cylinder leakage test the engine must be running to get proper results. Who is correct?
   A. Tech A
   B. Tech B
   C. Both A and B
   D. Neither A nor B

Engine Lubrication

Chapter Review

The following activities have been designed to help you refresh your knowledge of this chapter. Your instructor may require you to complete some or all of these activities as a regular part of your training program. You are encouraged to complete any activity that your instructor does not assign as a way to enhance your learning.

Matching

Match the following terms with the correct description or example.

A. Antifoaming agents

B. Bypass filter

C. Corrosion inhibitors

D. Crescent pump

E. Detergents

F. Dispersants

G. Extreme loading

H. Gelling

I. Hydro-cracking

J. Hydrogenating

K. Lubrication system

L. Oil cooler

M. Oil slinger

N. Oxidation inhibitor

O. Polyalphaolefin (PAO)

P. Rotor-type oil pump

Q. Scavenge pump

R. Synthetic blend

S. Viscosity index improver

T. Viscosity

_____ **1.** A thickening effect of oil in cold weather, wax content in base stock mineral oil makes it worse.

_____ **2.** A device used on small engines, located on the crankshaft or driven by the camshaft. It works to fling oil up onto moving engine parts.

_____ **3.** Oil additives that keep contaminants held in suspension in the oil, to be removed by the filter or when the oil is changed.

_____ **4.** The ability of a liquid to flow.

_____ **5.** A system of parts that work together to deliver lubricating oil to the various moving parts of the engine.

_____ **6.** A pump used with a dry sump oiling system to pull oil from the dry sump pan and move it to an oil tank outside the engine.

_____ **7.** Oil additives that help to keep carbon from sticking to engine components.

_____ **8.** A blend of conventional engine oil and pure synthetic oil.

_____ **9.** An oil filter system that only filters some of the oil.

_____ **10.** A process in which group 2 and group 3 oils are refined with hydrogen at much higher temperatures and pressures.

_____ **11.** An oil pump that uses a crescent-shaped part to separate the oil pump gears from each other, allowing oil to be moved from one side of the pump to the other.

_____ **12.** An oil additive that resists a change in viscosity over a range of temperatures.

_____ **13.** An oil additive that helps keep hot oil from combining with oxygen to produce sludge or tar.

_____ **14.** Oil additives that keep acid from forming in the oil.

_____ **15.** An oil pump that uses rounded gears to squeeze oil through.

_____ **16.** Oil additives that keep oil from foaming as it moves through the engine.

_____ **17.** A man-made base stock (synthetic) used in place of mineral oil. Oil molecules are more consistent in size, and no impurities are found in this oil since it is made in a lab.

_____ **18.** Large amount of pressure placed on two bearing surfaces to press oil from between them.

_____ **19.** A process used during refining of crude oil. Hydrogen is added to crude oil to create a chemical reaction to take out impurities such as sulfur.

_____ **20.** A device that takes heat away from engine oil by passing it near either engine coolant or outside air.

Multiple Choice

Read each item carefully, and then select the best response.

_____ **1.** What kind of oil varies in color from a dirty yellow to dark brown to black and can be thin like gasoline or a thick oil- or tarlike substance?
   A. Lubricating oil
   B. Motor oil
   C. Grease
   D. Crude oil

_____ **2.** _____ is a measure of how easily a liquid flows.
   A. Pour point
   B. Viscosity
   C. Solidity
   D. Reluctance

_____ **3.** What kind of additive coats parts with a protective layer so that the oil resists being forced out under heavy load?
   A. Extreme-pressure additives
   B. Viscosity index improver
   C. Pour point depressants
   D. Dispersants

_____ **4.** The _____ is the certifying body for engine oil.
   A. American Petroleum Institute
   B. American Society of Automotive Engineers
   C. International Lubricant Standardization and Approval Committee
   D. all of the above

_____ **5.** The letter _W_ in an SAE viscosity rating stands for _____.
   A. water content
   B. winter viscosity
   C. wax content
   D. weight

_____ **6.** Which of the following organizations sets classification standards for motorcycle engines, both two-stroke and four-stroke?
   A. American Society of Automotive Engineers
   B. Association des Constructeurs Européens d'Automobiles
   C. Japanese Automotive Standards Organization
   D. All of the above

_____ **7.** The American Petroleum Institute classifies oils into _____ groups.
   A. three
   B. four
   C. five
   D. six

_____ **8.** What type of oil can be man-made or highly processed petroleum?
   A. Group 1
   B. Synthetic
   C. Conventional
   D. Blended

_____ **9.** Lubrication oil is stored in the _____.
- **A.** oil slinger
- **B.** pickup tube
- **C.** oil gallery
- **D.** oil sump

_____ **10.** On a wet sump lubricating system, the _____ holds the entire volume of the oil required to lubricate the engine.
- **A.** drain plug
- **B.** oil pan
- **C.** crescents
- **D.** barriers

_____ **11.** In a _____ oil pump, the driving gear meshes with a second gear; as both gears turn, their teeth separate, creating a low-pressure area.
- **A.** geared
- **B.** crescent
- **C.** rotor-type
- **D.** spline-type

_____ **12.** The spin-on type oil filter has a(n) _____ that fits into a groove in the base of the filter.
- **A.** garter spring
- **B.** O-ring
- **C.** plastic cap
- **D.** compression ring

_____ **13.** Passageways called _____ allow oil to be fed to the crankshaft bearings first, then through holes drilled in the crankshaft to the connecting rods.
- **A.** pickup tubes
- **B.** sumps
- **C.** oil jackets
- **D.** galleries

_____ **14.** On horizontal-crankshaft engines, a _____ on the bottom of the connecting rod scoops up oil from the crankcase for the bearings.
- **A.** crescent pump
- **B.** dipper
- **C.** rotor
- **D.** strainer

_____ **15.** When the new O-ring in the new filter is installed over the old O-ring, is called _____.
- **A.** double-gasketing
- **B.** threaded filter
- **C.** reluctance
- **D.** hydro-cracking

True/False

If you believe the statement to be more true than false, write the letter "T" in the space provided. If you believe the statement to be more false than true, write the letter "F".

_____ **1.** Clearances, such as those between the crankshaft journal and crankshaft bearing, fill with lubricating oil so engine parts move or float on layers of oil instead of directly on each other.

_____ **2.** A power stroke can put as much as 2 tons of force on the main bearings.

_____ **3.** Oxidation allows air bubbles to form in the engine oil, reducing the lubrication quality of oil and contributing to the breakdown of the oil.

_____ **4.** The International Lubricant Standardization and Approval Committee requires that the oil provide increased fuel economy over a base lubricant.

_____ **5.** If you are servicing a European vehicle, it is advised that you do not go by any API recommendations; instead, make sure the oil meets the recommended JASO rating specified by the manufacturer.

_____ **6.** One of the impurities found in all crude oil is wax, which is removed during refining and is used for candle wax.

_____ **7.** Type 4 synthetic lubricating oil is not a true synthetic.

_____ **8.** On a wet sump lubricating system, the oil pan holds the entire volume of the oil required to lubricate the engine.

_____ **9.** The bypass filtering system is more common on diesel engines and is used in conjunction with a full-flow filtering system.

_____ **10.** The oil pressure sensor is also commonly called a sending unit since it sends a signal to the light, gauge, or message center in the dash.

_____ **11.** Oil analysis is typically used in heavy vehicle applications that may use 3 or more gallons of engine oil.

_____ **12.** Oil would be good for thousands of miles longer in a vehicle driven in stop-and-go traffic with long periods of idling than if it were driven in moderate temperatures for long distances.

_____ **13.** Because there is no oil storage sump under the engine, the engine can be mounted much lower in a dry sump system.

_____ **14.** Most small four-stroke gasoline engines use only splash lubrication to lubricate all of the parts on the engine.

_____ **15.** Checking the engine oil level regularly is necessary and should be performed during every fuel fill-up or every other fuel-up, depending on the age and condition of the vehicle.

Fill in the Blank

Read each item carefully, and then complete the statement by filling in the missing word(s).

1. _____ oil is distilled from crude oil and used as a base stock.

2. _____ occurs between all surfaces that come into contact with each other.

3. Base stock derived from crude oil will not retain its viscosity if the temperature gets cold enough, so viscosity _____ improvers are added to the stock.

4. The American _____ _____ symbol is the donut symbol located on the back of the oil bottle.

5. _____ oils flow easily during cold engine start-up but do not thin out as much as the engine and oil come up to operating temperature.

6. Using the wrong _____ can void the customer's warranty, leaving the customer, or your shop, responsible for repairs.

7. Group 2 and group 3 API classified oils are refined with _____ at much higher temperatures and pressures, in a process known as hydro-cracking.

8. Crude oil is broken down into _____ oil, which is then combined with additives to enhance the lubricating qualities.

9. True synthetic oils are based on man-made _____, commonly polyalphaolefin oil, which is a man-made oil base stock.

10. Oil is drawn through the oil pump _____ from the oil sump by an oil pump.

11. The oil pan is sealed to the engine with _____ or an oil pan gasket.

12. Between the oil sump and oil pump is a(n) _____ _____ with a flat cup and a wire mesh strainer immersed in the oil.

13. In a rotor-type oil pump, outside atmospheric pressure forces oil into the pump, and the oil fills the spaces between the _____ _____.

14. An oil pressure _____ _____ stops excess pressure from developing.

15. The location of the _____-_____ filter right after the oil pump ensures that all of the oil is filtered before it is sent to the lubricated components.

16. Some connecting rods have oil _____ holes that are positioned to receive oil from similar holes in the crankshaft.

17. Oil _____ systems are used to inform the driver when the oil needs to be changed.

18. Modern vehicle engines use a pressure, or _____-_____, lubrication system where the oil is forced throughout the engine under pressure.

19. Diesel fuel has more _____ thermal units of heat energy than gasoline, so it produces more heat when it is ignited, placing more stress on the engine's moving parts.

20. _____ can strip the threads on the oil pan, especially if it is made of aluminum, or can strip the threads on the drain plug itself.

Labeling

Label the following diagrams with the correct terms.

1. The lubrication system:

A. _____

B. _____

C. _____

D. _____

E. _____

F. _____

G. _____

H. _____

I. _____

2. Oil pump types:

A. _____

B

B. _____

C

C. _____

3. Full-flow filtering system:

B   C   D   E

F

H

A

G

A. _____

B. _____

C. _____

D. _____

E. _____

F. _____

G. _____

H. _____

4. Bypass filtering system:

A. _____

B. _____

C. _____

D. _____

E. _____

5. Tools for lubrication repair:

A. _____

B. _____

C. _____

D. _____

E. _____

F. _____

Crossword Puzzle

Use the clues to complete the puzzle.

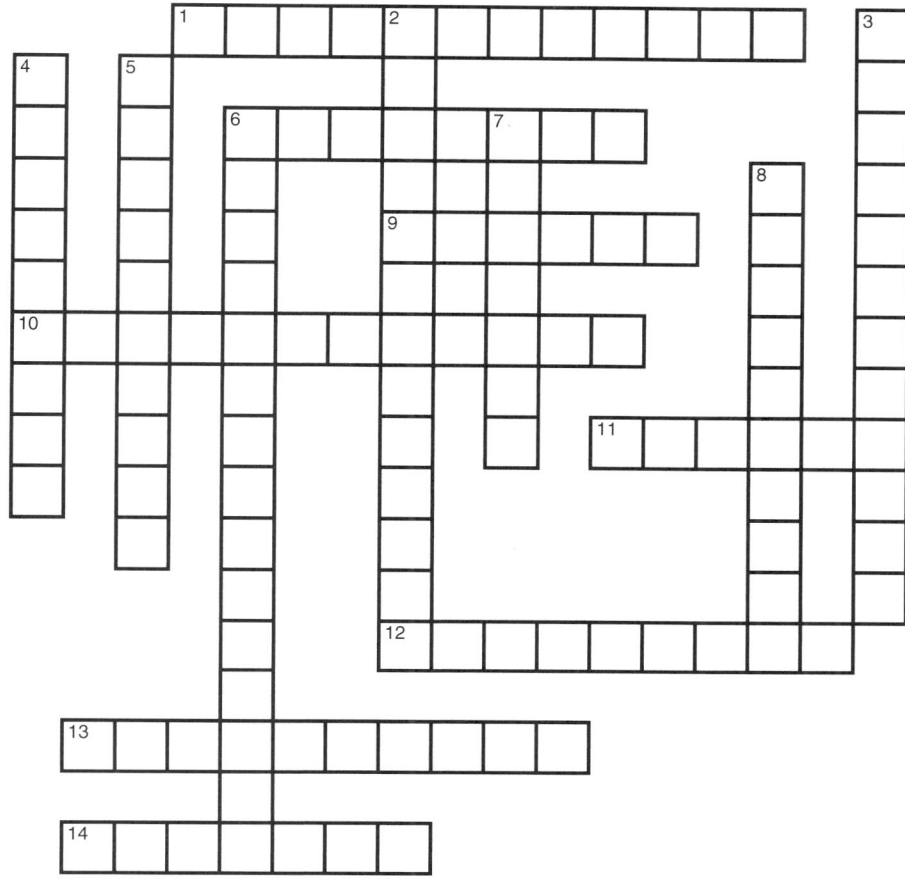

Across

1. This oil, in its pure form, uses man-made base stocks and is not derived from crude oil. This oil lasts longer and performs better than normal oil.
6. Material pulled from the earth, originating from organic compounds broken down over time and formed into petroleum.
9. The metal pan that covers the bottom of the engine, contains oil sump where engine oil is held.
10. A pump used with a dry sump oiling system to pull oil from the dry sump pan and move it to an oil tank outside the engine.
11. A type of splash lubricating system used in small engines. It works like a spoon scooping up oil and throwing it upward onto the crankshaft and other wear surfaces.
12. Passageways drilled or cast into the engine block or head(s), which carry pressurized lubricating oil to various moving parts in the engine, such as camshaft bearings.
13. Lobes or rounded edges on rotors that squeeze oil and create pressure.
14. A thickening effect of oil in cold weather.

Down

2. A process used during refining of crude oil. Hydrogen is added to crude oil to create a chemical reaction to take out impurities such as sulfur.
3. Oil passages that are drilled into the engine block and cylinder head(s). These passageways carry oil from the oil pump to critical moving parts.

4. The ability of a liquid to flow.

5. Base stock processed from crude oil in a refinery, used as the base material of all conventional oil.

6. Oil that is processed from crude oil; about 20% of oil is additives.

7. A device that pumps lubricating oil through the engine.

8. A tube connected to the oil pump that acts like a straw for the oil pump to pull oil from the sump of the oil pan.

ASE-Type Questions

Read each item carefully, and then select the best response.

_____ 1. Tech A says that one function of oil is to clean. Tech B says that one function of oil is to cushion. Who is correct?
  A. Tech A
  B. Tech B
  C. Both A and B
  D. Neither A nor B

_____ 2. Two techs are discussing 5W20 oil. Tech A says the W stands for "weight." Tech B says the W stands for "winter." Who is correct?
  A. Tech A
  B. Tech B
  C. Both A and B
  D. Neither A nor B

_____ 3. Tech A says that the higher the viscosity number, the thicker the oil. Tech B says that most modern vehicles use single weight oil. Who is correct?
  A. Tech A
  B. Tech B
  C. Both A and B
  D. Neither A nor B

_____ 4. Tech A says that spin-on oil filters need RTV gasket sealer to seal the gasket. Tech B says that some oil filters use a replaceable paper filter cartridge. Who is correct?
  A. Tech A
  B. Tech B
  C. Both A and B
  D. Neither A nor B

_____ 5. Tech A says that the oil pressure will typically be low if the oil level is at the "add" line on the dipstick. Tech B says that a cracked pickup tube could cause low oil pressure. Who is correct?
  A. Tech A
  B. Tech B
  C. Both A and B
  D. Neither A nor B

_____ 6. Tech A says that most oil pumps are of the positive displacement style. Tech B says that oil pumps are designed to deliver more oil than is needed for an engine. Who is correct?
  A. Tech A
  B. Tech B
  C. Both A and B
  D. Neither A nor B

_____ 7. Tech A says that it takes about 1 pint of oil to raise the oil level from "add" to "full." Tech B says that it takes about 1 quart to raise it that much. Who is correct?
  A. Tech A
  B. Tech B
  C. Both A and B
  D. Neither A nor B

_____ **8.** Tech A says that oil pressure is reduced when bearing clearances increase. Tech B says that oil pressure is regulated by the pressure relief valve. Who is correct?
  A. Tech A
  B. Tech B
  C. Both A and B
  D. Neither A nor B

_____ **9.** Tech A says that a full-flow oil filter filters all of the oil going to the bearings. Tech B says that a bypass filter bypasses the pump, so that any particles in the oil won't damage the pump. Who is correct?
  A. Tech A
  B. Tech B
  C. Both A and B
  D. Neither A nor B

_____ **10.** Tech A says that it is better to use two O-ring gaskets than just one on an oil filter. Tech B says that it is good practice to replace fiber or aluminum drain plug gaskets during every oil change. Who is correct?
  A. Tech A
  B. Tech B
  C. Both A and B
  D. Neither A nor B

Engine Cooling

Chapter Review

The following activities have been designed to help you refresh your knowledge of this chapter. Your instructor may require you to complete some or all of these activities as a regular part of your training program. You are encouraged to complete any activity that your instructor does not assign as a way to enhance your learning.

Matching

Match the following terms with the correct description or example.

A. Coolant
B. Coolant control valve
C. Cooling hoses
D. Cross-flow radiator
E. Down-flow radiator
F. Ethylene glycol
G. Heat dissipation
H. Overflow tank

I. Propylene glycol
J. Radiator
K. Radiator hoses
L. Radiation
M. Surge tank
N. Thermo-control switch
O. Thermostat
P. Water jackets

_____ **1.** A device that takes hot coolant and cools it by passing heat energy to the surrounding air.

_____ **2.** Flexible hoses that connect the stationary components of the cooling system such as the heater core and radiator to the engine, which is mounted on flexible mounts.

_____ **3.** Regulates the flow of coolant, allowing coolant to flow from the engine to the radiator when the engine is running at its operating temperature.

_____ **4.** The movement of energy through space such as the movement of energy from the sun to the earth.

_____ **5.** The passages in the engine block and cylinder head that surround the cylinders, valves, and ports.

_____ **6.** Rubber hoses that connect the radiator to the engine. Because they are subject to pressure, they are reinforced with a layer of fabric, typically nylon.

_____ **7.** A fluid that contains anti-freeze mixed with water.

_____ **8.** A temperature-sensitive switch that is mounted into the radiator or into a coolant passage on the engine to control electric fan operation.

_____ **9.** The spreading of heat over a large area to increase heat transfer.

_____ **10.** A valve that blocks off coolant flow to keep hot water from entering the heater core when less heat is requested by the operator.

_____ **11.** A chemical used as anti-freeze. It is labeled as a nontoxic anti-freeze.

_____ **12.** A radiator that uses cooling tubes that run horizontal with tanks on each end. This design allows lower hood profile for better vehicle aerodynamics.

_____ **13.** A tank used to catch any coolant that is released from the radiator cap (works like a catch can).

_____ **14.** A chemical used as anti-freeze that provides the lower freezing point of coolant and raises the boiling point. It is a toxic anti-freeze.

_____ **15.** A sealed tank that captures coolant coming from the head that has turned to steam and changes the steam back to coolant to be reused by the cooling system.

_____ **16.** A radiator that uses cooling tubes that run vertically. This design requires a higher hood profile.

Multiple Choice

Read each item carefully, and then select the best response.

_____ **1.** Heat transfers through liquids and gases by a process called _____.
- **A.** conduction
- **B.** convection
- **C.** radiation
- **D.** transference

_____ **2.** Engines have an ideal operating temperature somewhere around _____, give or take 20°F, depending on the vintage of the vehicle.
- **A.** 200°F
- **B.** 212°F
- **C.** 250°F
- **D.** 325°F

_____ **3.** The circulation of coolant is controlled by the _____.
- **A.** water pump
- **B.** radiator temperature sensor
- **C.** air temperature
- **D.** thermostat

_____ **4.** The best coolant is a _____ balance of water and anti-freeze, making it an ideal coolant for both hot and cold climates and providing adequate corrosion protection.
- **A.** 60/40
- **B.** 40/60
- **C.** 50/50
- **D.** 70/30

_____ **5.** A force pulling outward on a rotating body is known as _____.
- **A.** roll
- **B.** convection
- **C.** centrifugal force
- **D.** spin

_____ **6.** Coolant absorbs heat by _____ from the engine and becomes hotter.
- **A.** conduction
- **B.** convection
- **C.** radiation
- **D.** all of the above

_____ **7.** As coolant moves through the hottest part of an engine with a reverse-flow design, steam tends to form and get stuck at the head cooling passages; this problem was fixed by adding a(n) _____.
- **A.** overflow tank
- **B.** cooling fan
- **C.** surge tank
- **D.** recirculation pump

_____ **8.** What type of belt is used to drive the water pump and other accessories on the front of the engine?
- **A.** V-belt
- **B.** Serpentine belt
- **C.** Toothed belt
- **D.** Any of the above

_____ **9.** _____ are used around the radiator fan to help draw air through the entire radiator core and not just in front of the fan blades.
- **A.** Fins
- **B.** Baffles
- **C.** Shrouds
- **D.** Ducts

_____ **10.** The _____ is called a fan clutch since it can engage and disengage the cooling fan from the pulley.
   A. torque connector
   B. viscous coupler
   C. thermo-control switch
   D. electric solenoid

_____ **11.** The _____ is typically located on the water pump and connects to the intake manifold on many V-configured engines, such as a V6 or V8.
   A. bypass hose
   B. upper hose
   C. lower hose
   D. throttle body coolant line

_____ **12.** A _____ belt has a flat profile with a number of small V-shaped grooves running lengthwise along the inside of the belt.
   A. V-type
   B. serpentine
   C. stretch
   D. toothed

_____ **13.** The _____ operates by using a signal sent from a coolant temperature sensor located on the engine in a coolant passage.
   A. temperature gauge
   B. thermostat
   C. warning light
   D. both A and C

_____ **14.** The aluminum, brass, or steel plugs designed to seal the openings left from the casting process where the casting sand was removed are called _____.
   A. core plugs
   B. soft plugs
   C. expansion plugs
   D. all of the above

_____ **15.** What type of coolant contains a mixture of inorganic and organic additives?
   A. Organic acid technology
   B. Inorganic acid technology
   C. Hybrid organic acid technology
   D. Poly organic acid technology

True/False

If you believe the statement to be more true than false, write the letter "T" in the space provided. If you believe the statement to be more false than true, write the letter "F".

_____ **1.** Heat is transferred from one solid to another by a process called radiation.

_____ **2.** No matter how efficiently fuel burning occurs, and no matter the size of the engine, the heat energy generated never completely transforms into kinetic energy.

_____ **3.** Some newer vehicles have a coolant heat storage system that uses a vacuum-insulated container similar to a Thermos bottle.

_____ **4.** Water alone is by far the best coolant there is, since it can absorb a larger amount of heat than most other liquids.

_____ **5.** Ethylene glycol is a chemical that resists freezing but is not toxic and is used in nontoxic anti-freezes.

_____ **6.** Because atmospheric pressure is lower at higher elevations, the boiling temperature of a liquid in an unsealed system is higher.

_____ **7.** In an automotive cooling system, electrolysis is possible when the coolant breaks down and becomes more acidic.

_____ **8.** Cooling systems, such as those used for turbocharger intercoolers/aftercoolers or hybrid high-voltage battery packs, may use air-to-air systems to cool the engine's intake air or high-voltage battery.

_____ **9.** All automobiles are now water cooled, although small engines can still be found in older automobiles that are air cooled.

_____ **10.** All engines operate best when they are at their full operating temperature.

_____ **11.** The hottest part of any engine is the cylinder head, since this is where the combustion chamber is located.

_____ **12.** Radiator hoses connect the water pump and engine to the heater core. They carry heated coolant to the heater core to be used to heat the passenger compartment.

_____ **13.** In both the cross-flow and the down-flow radiator designs, the core is built of the same components.

_____ **14.** The thermostat is a spring-loaded valve that is controlled by a wax pellet located inside the valve.

_____ **15.** Thermo-control switches often use a bimetallic strip that consists of two different metals or alloys laminated back to back that expand and contract at different rates.

_____ **16.** Some radiator hoses, especially lower hoses, have a spiral wire inside to keep the hose from collapsing during heavy acceleration when the water pump is drawing a lot of water from the radiator.

_____ **17.** The bottom radiator hose is typically attached to the thermostat housing, which allows the heated coolant to enter the inlet side of the radiator.

_____ **18.** The heater control valve, if used, controls the flow of coolant to the heater core to control the temperature of the air desired by the operator.

_____ **19.** Coolant can be easily identified according to its color, which may be anything from green or purple to yellow/gold, orange, blue, or pink.

_____ **20.** Poly organic acid technology coolant contains a proprietary blend of corrosion inhibitors and provides up to 7 years or 250,000 miles of protection.

Fill in the Blank

Read each item carefully, and then complete the statement by filling in the missing word(s).

1. Heat is _____ energy. It cannot be destroyed; it can only be transferred.

2. Heat is transferred through space by a process called _____.

3. The principle that heat always moves from hot to cold is known as the second law of _____.

4. Coolant is a mixture of water and _____-_____, which is used to remove heat from the engine.

5. _____ glycol is a chemical that resists freezing but is very toxic to humans and animals.

6. A liquid under pressure higher than atmospheric pressure has a _____ boiling point than when that liquid is at atmospheric pressure.

7. _____ engine cooling uses heat-dissipating fins on the engine cylinders and heads to allow the movement of air to absorb heat and carry it away from the engine through special ducting.

8. In a typical _____-_____ design, coolant starts from the radiator and flows through the radiator outlet hose to the thermostat and then to the water pump.

9. The _____ is usually made of copper, brass, or aluminum tubes with copper, brass, aluminum, or plastic tanks on the sides or top for coolant to collect in.

10. _____ coolers are used to cool automatic transmission fluid, power steering fluid, EGR gasses, and compressed intake air.

11. One way to prevent coolant from boiling is to use a radiator _____ cap.

12. The coolant _____ system consists of an overflow bottle, a sealed radiator pressure cap, and a small hose connecting the bottle to the radiator neck.

13. The _____ _____ is usually belt driven from a pulley on the front of the crankshaft.

14. Engine-driven _____ _____ may be located on the water pump shaft, or in a few cases it may be attached directly to the engine crankshaft.

15. The bottom or lower _____ _____ is connected between the outlet of the radiator and the inlet of the water pump.

16. A(n) _____ belt looks like an ordinary serpentine belt but is found on vehicles without a tensioner.

17. _____ are used to keep the drive belt tight around the pulleys to ensure the least amount of slippage without causing damage to component bearings.

18. Temperature _____ can come in two forms: a temperature gauge or a temperature light located in the instrument cluster.

19. The _____ _____ is simply a small radiator that is mounted inside the heater box in the passenger compartment.

20. Air doors are moved by one of three methods: cable, vacuum actuator, or an electric _____, called a stepper motor.

Labeling

Label the following diagrams with the correct terms.

1. Heat transfer:

A. _____

B. _____

C. _____

D. _____

2. Percent of heat loss in a gasoline engine:

A. _____

B. _____

C. _____

D. _____

3. Types of radiators:

A. _____

B. _____

4. Coolant recovery system:

A. _____

B. _____

C. _____

D. _____

5. Thermostat valve:

A. _____

B. _____

6. Water pump:

A. _____

B. _____

C. _____

D. _____

7. Hydraulically operated cooling fan:

A. _____

B. _____

C. _____

D. _____

8. Cooling system:

A. _____

B. _____

C. _____

D. _____

E. _____

Skill Drills

Test your knowledge of skill drills by filling in the correct words in the photo captions.

    1. Testing the Cooling System Pressure:

Step 1: Inspect for _____. Verify specified cooling system _____, install the radiator _____ on the _____ tester, and pressurize the cap to the correct pressure. It should hold pressure at approximately the _____ pressure and _____ at slightly above the rated pressure.

Step 2: Top off the radiator with _____ or _____, and install the tester. _____ the system to the _____ cap pressure.

Step 3: Watch the pressure reading for a _____ while performing a _____ check for any leaks. Check heater _____, soft _____, and any heater _____; determine necessary action.

2. Using a Hydrometer to Test the Freeze Point of the Coolant:

Step 1: Remove the pressure cap. Place the _____ tube in the _____, and _____ the ball on top.

Minimum Level

Step 2: _____ the ball to pull a coolant _____ into the hydrometer. _____ it is above the _____ level in the tester.

Step 3: Read the _____ on the tool to verify the _____ _____ of the coolant. Return coolant sample to the _____ or _____ tank.

3. Draining and Refilling Coolant:

Step 1: Locate the radiator _____ _____, if equipped, and place a catch _____ marked for coolant underneath the _____ _____. Drain the radiator into the catch pan.

Step 2: Remove the _____ drain plugs, and allow the _____ to drain into the pan.

Step 3: Refill the cooling _____ with the proper coolant _____ after _____ the drain plugs. _____ the engine, and _____ the proper level. _____ of coolant in an _____ way.

4. Checking and Replacing a Coolant Hose:

Step 1: Inspect the hoses by _____ them, and _____ inspect the _____. Remove the _____ _____. Carefully pull or _____ the hose off the fittings.

Step 2: Verify that the _____ hose is correct. Cut to _____ if necessary. Reinstall the new _____ all the way into position. Install new _____.

Step 3: Refill _____, verify correct _____, and _____ check for leaks.

Crossword Puzzle

Use the clues to complete the puzzle.

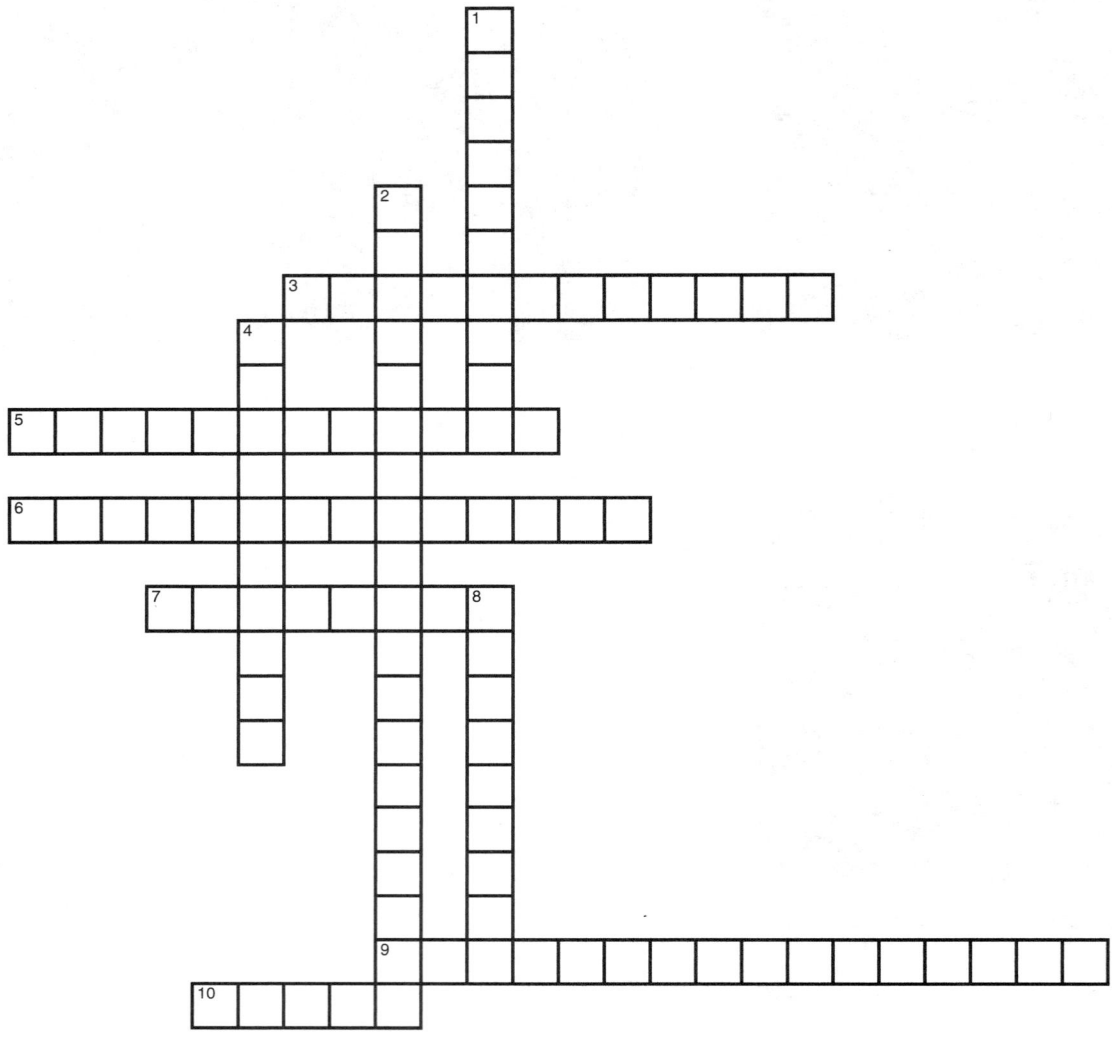

Across

3. The process of pulling metals apart by using electricity or by creating electricity through the use of chemicals and dissimilar metals.

5. The temperature at which a substance begins to change from a liquid to a gas.

6. Called a fan clutch, a hub that connects the water pump drive to the cooling fan uses a temperature-sensitive viscous fluid to cause the fan to turn faster as the temperature of the air pulled through the radiator increases.

7. A device that is electrically or vacuum controlled and is used to physically move doors within the heater box to control airflow.

9. A force pulling outward on a rotating body.

10. The angle of a fan blade. A steeper pitch draws more air, while a shallower pitch draws less air.

Down

1. Movement of heat energy through gases or liquids.

2. A temperature-sensitive switch that is mounted into the radiator or into a coolant passage on the engine to control fan operation.

4. Movement of heat energy through solids.

8. The movement of energy through space such as the movement of energy from the sun to the earth.

ASE-Type Questions

Read each item carefully, and then select the best response.

_____ 1. Tech A says that the cooling system is designed to keep the engine as cool as possible. Tech B says that the heater core can remove heat from the cooling system. Who is correct?
   A. Tech A
   B. Tech B
   C. Both A and B
   D. Neither A nor B

_____ 2. Tech A says that the thermostat is open until the engine warms up, and then it closes. Tech B says that a faulty radiator cap can be the cause of boiling coolant. Who is correct?
   A. Tech A
   B. Tech B
   C. Both A and B
   D. Neither A nor B

_____ 3. Tech A says that when pure anti-freeze is used in the cooling system, the protection level is –70°F. Tech B says that pure anti-freeze will cool the engine better than water. Who is correct?
   A. Tech A
   B. Tech B
   C. Both A and B
   D. Neither A nor B

_____ 4. Tech A says that some drive belts are a stretch fit design. Tech B says that the thermostat may have a bleed valve that should be accurately positioned when the thermostat is replaced. Who is correct?
   A. Tech A
   B. Tech B
   C. Both A and B
   D. Neither A nor B

_____ 5. Tech A says that the design of a surge tank system helps to purge air from the cooling system. Tech B says that overflow tanks are pressurized. Who is correct?
   A. Tech A
   B. Tech B
   C. Both A and B
   D. Neither A nor B

_____ 6. Tech A says that when you find coolant hoses collapsed, the radiator pressure cap has failed. Tech B says that overflow tanks are reservoirs designed to hold coolant and allow flow back to the radiator. Who is correct?
   A. Tech A
   B. Tech B
   C. Both A and B
   D. Neither A nor B

_____ 7. Tech A says that electric cooling fans operate whenever the engine is running. Tech B says that electric cooling fans are used to cause a large airflow over the radiator at low vehicle speeds. Who is correct?
   A. Tech A
   B. Tech B
   C. Both A and B
   D. Neither A nor B

_____ 8. Tech A says that there are a number of coolant types that each has its own life span. Tech B says that mixing of coolants is generally ok, since all coolants use the same chemical base. Who is correct?
   A. Tech A
   B. Tech B
   C. Both A and B
   D. Neither A nor B

_____ **9.** Tech A says that coolant leaking out of the water pump weep hole is normal. Tech B says that a refractometer measures the specific gravity of a liquid. Who is correct?

    A. Tech A

    B. Tech B

    C. Both A and B

    D. Neither A nor B

_____ **10.** Tech A says that a cooling system pressure tester can be used to find leaks in the cooling system. Tech B says that the cooling system pressure tester can be used to test the pressure at which the radiator cap vents. Who is correct?

    A. Tech A

    B. Tech B

    C. Both A and B

    D. Neither A nor B

Automatic Transmission Fundamentals

Chapter Review

The following activities have been designed to help you refresh your knowledge of this chapter. Your instructor may require you to complete some or all of these activities as a regular part of your *Fundamentals of Automotive Technology* training program. You are encouraged to complete any activity that your instructor does not assign as a way to enhance your learning in the classroom.

Matching

Match the following terms with the correct description or example.

A. Extension housing
B. Fluid coupler
C. Gear ratio
D. Helical-cut gear
E. One-way clutch

F. Planetary gears
G. Ring gear
H. Spur gear
I. Thrust bearing
J. Transaxle

_____ 1. A type of gear in which the teeth of the gear are cut in a straight line down the axis of the gear.

_____ 2. A type of hydraulic coupling used on vintage vehicles to connect and transfer power from the engine to the transmission.

_____ 3. A type of gear in which the teeth are cut in a spiral down the axis of the gear.

_____ 4. Also called a Torrington bearing, a small roller bearing assembly with the rollers laid flat axially around the centerline of the bearing. The bearing is used to control forward and backward movement of a part in an automatic transmission.

_____ 5. The small gears in a planetary gear set that revolve around the sun gear; also known as pinion gears.

_____ 6. A component of the automatic transmission housing that covers the output shaft of the transmission. It also supports the end of the driveshaft and may hold components such as the vehicle speed sensor, speedometer drive assembly, and governor assembly.

_____ 7. The outer gear of a planetary gear set that wraps around the outside of the planetary gears.

_____ 8. The ratio of the size or teeth of one gear compared to the size or teeth of a mating gear.

_____ 9. A type of transmission typically used in front-wheel drive vehicles in which the transmission also includes the differential and final drive gear.

_____ 10. A type of holding device used by an automatic transmission to stop the movement of one component of a planetary gear set. It allows free spinning in one direction but will lock up when the part attempts to spin in the opposite direction.

Multiple Choice

Read each item carefully, and then select the best response.

_____ **1.** Transmissions that use two pulleys that change diameter in response to vehicle load and speed, are known as_____.
   A. dual-clutch transmissions
   B. continuously variable transmissions
   C. power-splitting transmissions
   D. automatic transmissions

_____ **2.** With which type of hybrid drive train does the engine have two or more methods for the power to flow through the transmission?
   A. Series hybrid
   B. Parallel hybrid
   C. Dual-clutch
   D. Continuously variable

_____ **3.** Early automatic transmissions used a _____ rather than a torque converter.
   A. fluid coupler
   B. flywheel
   C. transaxle
   D. countershaft

_____ **4.** In its simplest form, a single-stage torque converter has three elements including all of the following *except* _____.
   A. turbine
   B. impeller
   C. fluid coupler
   D. stator

_____ **5.** The _____ has a small set of curved blades attached to a central hub and is positioned between the impeller and the turbine.
   A. fluid coupler
   B. pinion
   C. servo
   D. stator

_____ **6.** Automatic transmission vehicles use a _____, sometimes called a transmission cooler, in one tank of the radiator.
   A. evaporator
   B. heat exchanger
   C. blend box
   D. condenser

_____ **7.** The difference in diameter between the driving gear and the driven gear is known as _____.
   A. leverage
   B. torque conversion
   C. gear ratio
   D. reduction

_____ **8.** What type of gears are often used to change the direction of power flow by 90 degrees?
   A. Helical-cut gears
   B. Sun gears
   C. Planetary gears
   D. Hypoid gears

_____ **9.** A friction-lined steel belt that wraps around the outside of a drum inside the automatic transmission is called a _____.
   A. band
   B. clutch
   C. pinion
   D. strap

_____ **10.** Which of the following should never be used to lubricate an internal transmission seal?
    A. ATF fluid
    B. Grease
    C. Petroleum jelly
    D. Automatic transmission assembly lubricant

True/False

If you believe the statement to be more true than false, write the letter "T" in the space provided. If you believe the statement to be more false than true, write the letter "F".

_____ **1.** An automatic transmission can select and shift gears without input from the driver.

_____ **2.** In a manual transmission, engine torque is controlled by two, typically wet, clutches.

_____ **3.** Modern transmissions no longer use fluid couplers.

_____ **4.** The stator has a small set of curved blades attached to a central hub and is positioned between the impeller and the turbine.

_____ **5.** Most automatic transmission vehicles use a heat exchanger, sometimes called a transmission cooler, in one tank of the radiator.

_____ **6.** Planetary gears are held in place in a planet carrier.

_____ **7.** Helical-cut gears tend to be much louder and do not offer as much strength as the other types of gears.

_____ **8.** Automatic transmission fluids are typically dyed purple for easy identification.

_____ **9.** On a manual transmission there is no flywheel; there is a thin, lightweight steel flex plate.

_____ **10.** Teflon seals come in several styles—continuous, butt-cut, scarf-cut, and step joint.

Fill in the Blank

Read each item carefully, and then complete the statement by filling in the missing word(s).

1. _____-_____ transmissions are a newer type of automatic transmission that uses two, typically wet, clutches.

2. In a _____ hybrid drive train, the electric motor is not typically able to propel the vehicle on its own. This electric motor is often placed between the engine and the transmission.

3. A series-parallel hybrid drive train uses what is called a _____-_____ transmission.

4. The _____ _____ is mounted between the engine and the transmission, in the same place as a manual transmission clutch.

5. A _____ _____ is basically two fans facing each other.

6. The _____ has a large number of vanes attached to the torque converter housing to form the driving member.

7. Planetary gears revolve inside a larger _____ gear that wraps around the outside of the whole planetary gear set.

8. _____ gears were used in early transmissions due to ease of manufacturing and lower cost.

9. One-way clutches can be either one-way _____ or _____.

10. When the driver shifts the transmission into park, a lever, called a _____ _____, is forced into notches cut into a hardened steel drum on the output shaft of the transmission.

Labeling

Label the following diagrams with the correct terms.

1. Dual-clutch transmission:

A. _____

B. _____

C. _____

D. _____

E. _____

F. _____

G. _____

H. _____

I. _____

J. _____

K. _____

L. _____

M. _____

2. Torque converter mounted between the engine and the transmission:

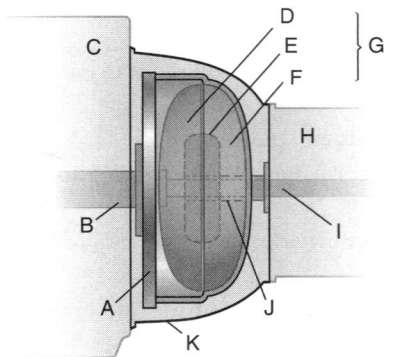

A. _____

B. _____

C. _____

D. _____

E. _____

F. _____

G. _____

H. _____

I. _____

J. _____

K. _____

3. Fluid flow through a torque converter:

A. _____

B. _____

C. _____

D. _____

E. _____

F. _____

G. _____

H. _____

I. _____

J. _____

4. Simple planetary gear set:

A. _____

B. _____

C. _____

D. _____

5. Gear set styles:

A. _____

B. _____

C. _____

6. A hypoid gear arrangement:

A. _____

B. _____

7. Typical transmission band:

A. _____

B. _____

C. _____

D. _____

E. _____

F. _____

G. _____

Crossword Puzzle

Use the clues in the column to complete the puzzle.

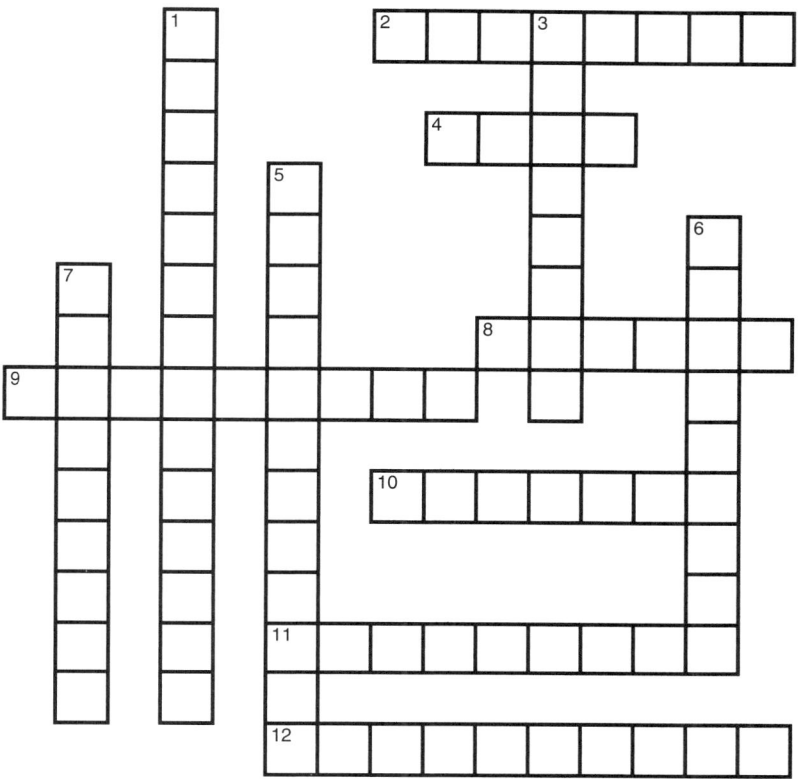

Across

2. A type of gear in which the teeth of the gear are cut in a straight line down the axis of the gear.

4. A metal band with friction material bonded to one side. The band is contracted around a drum to stop the drum from spinning.

8. A rubber, cork, or paper spacer that goes between two parts to seal the gap between the parts.

9. The ratio of the size or teeth of one gear compared to the size or teeth of a mating gear.

10. The center gear of a planetary gear seat around which the other gears rotate.

11. A type of transmission typically used in front-wheel drive vehicles in which the transmission also includes the differential and final drive gear.

12. A type of helical gear used to change the direction of motion 90 degrees. The axis of the input gear does not line up on the centerline of the output gear.

Down

1. The small gears in a planetary gear set that revolve around the sun gear.

3. The outer gear of a planetary gear set. It wraps around the outside of the planetary gears.

5. A type of holding device used by an automatic transmission to stop the movement of one component of a planetary gear set. It allows free spinning in one direction but will lock up when the part attempts to spin in the opposite direction.

6. Any gear ratio that results in a torque reduction with a speed increase. Overdrive is used on vehicles to reduce the engine speed when traveling at highway speed in order to save fuel.

7. Typically a circular steel plate that is bolted to the rear of the crankshaft.

ASE-Type Questions

Read each item carefully, and then select the best response.

_____ 1. Tech A says that the higher the numerical gear ratio (4:1), the more torque that will be applied to the wheels. Tech B says that the lower the numerical gear ratio (2:1), the more torque that will be applied to the wheels. Who is correct?
   A. Tech A
   B. Tech B
   C. Both A and B
   D. Neither A nor B

_____ 2. Tech A says that helical-cut gears are stronger than straight-cut gears. Technician B says that helical-cut gears are noisier than straight-cut gears. Who is correct?
   A. Tech A
   B. Tech B
   C. Both A and B
   D. Neither A nor B

_____ 3. Tech A says that a transaxle must have a type of differential assembly to allow the front wheels to turn at different speeds while driving around a curve. Tech B says that a conventional rear-wheel drive vehicle will typically use a differential assembly that is located in the rear axle assembly. Who is correct?
   A. Tech A
   B. Tech B
   C. Both A and B
   D. Neither A nor B

_____ 4. Tech A says that a parallel hybrid transmission has two or more different devices to propel the vehicle. Tech B says that hybrid power-splitting transmissions use two adjustable pulleys and a heavy metal belt. Who is correct?
   A. Tech A
   B. Tech B
   C. Both A and B
   D. Neither A nor B

_____ 5. Tech A says that planetary gear sets are used in all automatic transmissions. Tech B says that planetary gear sets are often combined together to create the needed gear ratios for a transmission. Who is correct?
   A. Tech A
   B. Tech B
   C. Both A and B
   D. Neither A nor B

_____ 6. Tech A says that the impeller is connected directly to the flex plate. Tech B says that the turbine shaft drives the front pump. Who is correct?
   A. Tech A
   B. Tech B
   C. Both A and B
   D. Neither A nor B

_____ 7. Tech A says that most automatic transmissions have a ring gear for the starter as part of the flex plate. Tech B says that the ring gear is typically replaceable (interference fit) on a flex plate. Who is correct?
   A. Tech A
   B. Tech B
   C. Both A and B
   D. Neither A nor B

_____ 8. Tech A says that a knocking noise in the engine/transmission area is typically a bad front pump in the transmission. Tech B says that a knocking noise in the engine/transmission area is more likely to be loose flex plate bolts. Who is correct?
   A. Tech A
   B. Tech B
   C. Both A and B
   D. Neither A nor B

_____ **9.** Tech A says that painting a rebuilt transmission helps to identify it as being a rebuilt transmission. Tech B says that painting a rebuilt transmission prevents the housing from leaking due to case porosity. Who is correct?

    A. Tech A

    B. Tech B

    C. Both A and B

    D. Neither A nor B

_____ **10.** Tech A says that the torque converter multiplies the amount of torque transmitted from the engine to the transmission. Tech B says that the torque converter couples and uncouples the engine and transmission as the vehicle stops and starts in traffic. Who is correct?

    A. Tech A

    B. Tech B

    C. Both A and B

    D. Neither A nor B

Servicing the Automatic Transmission/Transaxle

Chapter Review

The following activities have been designed to help you refresh your knowledge of this chapter. Your instructor may require you to complete some or all of these activities as a regular part of your *Fundamentals of Automotive Technology* training program. You are encouraged to complete any activity that your instructor does not assign as a way to enhance your learning in the classroom.

Matching

Match the following terms with the correct description or example.

A. Bidirectional control

B. Cooler flow test

C. Hydraulic pressure test

D. Noise, vibration, and harshness (NVH) test

E. Power train mount

F. Stall test

_____ 1. A test to measure for any audible noises, vibrations, and harsh operation. It can be completed by the technician with or without the aid of a tester. The tester is used to pinpoint the exact frequencies of the noise and vibrations.

_____ 2. A rubber or metal bracket used to secure the engine and transmission into the vehicle. Some vehicles use hydraulic or electrohydraulic power train mounts.

_____ 3. The ability to command different solenoids and actuators "on" and "off" to check their operation.

_____ 4. A test that involves raising the engine rpm to wide-open throttle while the brake is firmly applied and the transmission is in gear. It is used to check torque convertor and transmission operation on some vehicles.

_____ 5. The placement of a specialty measuring device into the transmission cooler line to measure fluid flow to the cooler.

_____ 6. The use of a hydraulic pressure gauge to measure the amount of hydraulic pressure produced in each gear range.

Multiple Choice

Read each item carefully, and then select the best response.

_____ 1. How much fluid does it take to raise the level from the bottom of the crosshatched area or add mark to the full mark on most transmission dipsticks?

A. Half pint

B. Quart

C. Pint

D. Gallon

_____ 2. When raised in the air on a hoist, a vehicle's speed should never be allowed to exceed _____ mph on the speedometer.

A. 10

B. 25

C. 35

D. 45

_____ **3.** What type of test checks the pump operation, the pressure regulator, and the seals and gaskets inside the transmission?
  A. Air check
  B. NVH test
  C. Hydraulic pressure test
  D. Cooler flow test

_____ **4.** What type of test checks torque converter operation and also checks the transmission for slippage by operating the engine at full throttle with the transmission in gear and the brakes applied?
  A. Hydraulic pressure test
  B. Stall test
  C. NVH test
  D. Preload test

_____ **5.** A test tool that can help pinpoint noises in a vehicle while the vehicle is driven, called a _____.
  A. NVH analyzer
  B. stethoscope
  C. scan tool
  D. PID tester

_____ **6.** What feature makes a high-quality aftermarket scan tool or factory scan tool better than a code reader?
  A. Live data
  B. Reads DTCs
  C. Bidirectional control
  D. Both A and C

_____ **7.** Which of the following repairs can be performed without the removal of the transmission?
  A. Replace shift solenoid
  B. Replace speed sensor
  C. Replace power train mounts
  D. All of the above

_____ **8.** Replacement of which of the following requires the removal of the driveshaft and the extension housing?
  A. Speed sensor seal
  B. Extension housing bushing
  C. Pan gasket
  D. All of the above

_____ **9.** A loud thump when the accelerator is applied in drive or reverse and again when the brake pedal is applied may indicate a _____.
  A. broken power train mount
  B. dry extension-housing bushing
  C. transmission fluid leak
  D. bad shift sensor

_____ **10.** If the _____ is not adjusted properly, the transmission position indicator will not be set correctly and the customer will not know which gear the vehicle is in.
  A. speed sensor
  B. shift sensor
  C. shift linkage
  D. lock-up torque converter

True/False

If you believe the statement to be more true than false, write the letter "T" in the space provided. If you believe the statement to be more false than true, write the letter "F".

_____ **1.** Some vehicles do not have a transmission dipstick.

_____ **2.** All transmissions are checked with the engine at operating temperature, idling, and the transmission in park.

_____ **3.** Some modern transmission fluids are a darker red when they are brand new and even have a slightly burnt smell to them.

_____ **4.** The presence of a small amount of clutch material or metal in the bottom of the pan indicates a serious transmission problem.

_____ **5.** Scanning the power train control module for trouble codes will indicate the specific transmission problem.

_____ **6.** Because a stall test stresses the drive train so much, many manufacturers do not recommend performing the test anymore.

_____ **7.** During test-drive, it is critical to place the transmission in the same operating conditions as the customer stated the problem occurred in.

_____ **8.** Bidirectional control allows the technician to command the chassis dynamometer to operate in "forward" or "reverse" without moving the shift selector in the vehicle.

_____ **9.** If the manual valve shift linkage is far enough out of adjustment, the valve may supply fluid to the wrong gear.

_____ **10.** If a transmission suffers catastrophic failure, metal particles and old clutch material can become stuck inside the fluid cooler or lines.

Fill in the Blank

Read each item carefully, and then complete the statement by filling in the missing word(s).

1. When fluids leak, they travel _____ due to gravity and typically toward the _____ of the vehicle.

2. If a transmission leak is hard to locate, place a leak detection _____ in the transmission fluid.

3. The new way to check the quality of transmission fluid is to take a few drops and place them on a clean _____ _____.

4. During the _____-_____, it is critical to place the transmission in the same operating conditions as the customer stated the problem occurred in.

5. Many high-quality or factory scan tools give the technician _____ control of the transmission, allowing them to command different solenoids and actuators "on" and "off" to check their operation.

6. The power train _____ hold the engine and transmission in the proper position in the vehicle.

7. With worn power train mounts, the _____ _____ can become bound, and the driver might be able to start a vehicle while it is in gear or have the vehicle in the wrong gear.

8. Most power train mounts are of the _____ style and subject to cracks and tears that occur over time due to their constant flexing.

9. Some _____ mounts use a computer-controlled electronic control valve that can open an alternate passageway or vary the size of the orifice.

10. A transmission fluid _____ that is leaking could cause the new transmission to fail from low fluid level or could cause antifreeze to be drawn into the cooler from the radiator, contaminating the fluid.

Crossword Puzzle

Use the clues in the column to complete the puzzle.

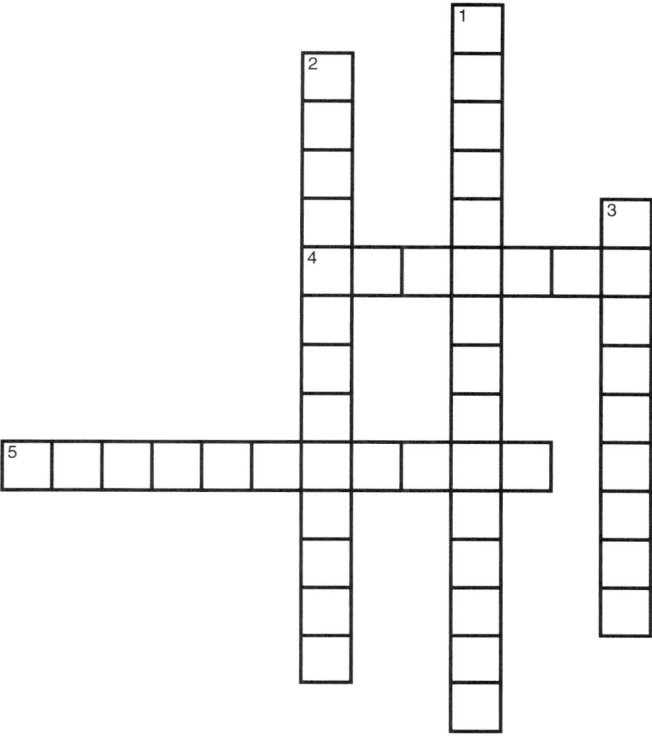

Across

4. A test to measure any audible noises, vibrations, and harsh operation.

5. The use of compressed air to check clutch and servo operation on transmissions during and after assembly.

Down

1. A rubber or metal bracket used to secure the engine and transmission into the vehicle.

2. The process of removing the transmission pan to check for clutch material, metal, and other debris or contaminants.

3. A test that involves raising the engine rpm to wide-open throttle while the brake is firmly applied and the transmission is in gear. The test is used to check torque convertor and transmission operation on some vehicles.

ASE-Type Questions

Read each item carefully, and then select the best response.

_____ 1. Tech A says that you should check fluid level in many automatic transmissions with the engine idling and the transmission in park. Tech B says that some transmissions are made without dipsticks. Who is correct?
   A. Tech A
   B. Tech B
   C. Both A and B
   D. Neither A nor B

_____ 2. Tech A says that increasing the transmission fluid level from the add mark to the full mark will require a quart of fluid, just like with engine oil. Tech B says that transmission solenoids should be tested electrically and mechanically. Who is correct?
   A. Tech A
   B. Tech B
   C. Both A and B
   D. Neither A nor B

_____ 3. Tech A says that transmission pan gaskets are always replaced. Tech B says that some newer transmissions have reusable pan gaskets. Who is correct?
   A. Tech A
   B. Tech B
   C. Both A and B
   D. Neither A nor B

_____ 4. Tech A says that the cooler and cooler lines should be flushed as part of a transmission rebuild. Tech B says that when checking the fluid level on the dipstick, always read the highest level on either side. Who is correct?
   A. Tech A
   B. Tech B
   C. Both A and B
   D. Neither A nor B

_____ 5. Tech A says that using a scan tool to activate the TCC with the engine idling in drive and the brakes applied is a typical troubleshooting task. Tech B says that diagnosis is a waste of time on a faulty transmission since it will be rebuilt anyway. Who is correct?
   A. Tech A
   B. Tech B
   C. Both A and B
   D. Neither A nor B

_____ 6. Tech A says that a stall test checks the torque-converter operation and also checks the transmission for slippage. Tech B says the stall test stresses the torque converter and transmission, so any weaknesses will be shown as a result. Who is correct?
   A. Tech A
   B. Tech B
   C. Both A and B
   D. Neither A nor B

_____ 7. Tech A says that if the linkage is not adjusted properly, the transmission position indicator (PRNDL indicator) will not be set correctly and the driver may not know which gear the vehicle is in. Tech B says that even if the linkage is not adjusted properly, the backup or reverse lights will still work normally. Who is correct?
   A. Tech A
   B. Tech B
   C. Both A and B
   D. Neither A nor B

_____ **8.** Tech A says you should check fluid level in all types of automatic transmissions with the engine idling and the transmission in park. Tech B says that a transmission fluid leak can result in failure of the transmission due to a low fluid level. Who is correct?

A. Tech A

B. Tech B

C. Both A and B

D. Neither A nor B

_____ **9.** Tech A says that often in the event of a broken power train mount, customers complain of a loud thump when they apply the accelerator in drive or reverse. Tech B says that the loud thump may be the engine and/or transmission being pulled up from the broken mount due to the engine's torque, and then being set back down when the brake is applied. Who is correct?

A. Tech A

B. Tech B

C. Both A and B

D. Neither A nor B

_____ **10.** Tech A says that a transmission fluid cooler that is leaking could cause the new transmission to fail from low fluid level. Tech B says that such a leak could cause antifreeze to be drawn into the cooler from the radiator and contaminate the transmission fluid. Who is correct?

A. Tech A

B. Tech B

C. Both A and B

D. Neither A nor B

Hybrid and Continuously Variable Transmissions

Chapter Review

The following activities have been designed to help you refresh your knowledge of this chapter. Your instructor may require you to complete some or all of these activities as a regular part of your *Fundamentals of Automotive Technology* training program. You are encouraged to complete any activity that your instructor does not assign as a way to enhance your learning in the classroom.

Matching

Match the following terms with the correct description or example.

A. Belt alternator starter (BAS)
B. Continuously variable transmission (CVT)
C. Electronic continuously variable transmission (ECVT)
D. Hybrid drive system
E. Integrated motor assist (IMA)
F. Regenerative braking
G. Torque assist
H. Torque smoothing
I. Variable-diameter pulley (VDP)

_____ 1. A type of CVT that uses two pulleys with moveable sheaves, allowing the effective diameter of the pulleys to change, resulting in variable gear ratios.

_____ 2. A drive system that uses two or more propulsion systems such as electric motors and an internal combustion engine (ICE).

_____ 3. A type of hybrid drive system that uses a belt-driven alternator/starter that operates on 42 volts.

_____ 4. A process that uses an electric motor to smooth out engine power pulses when an ICE is operating at low rpm or when the vehicle is using fuel management techniques such as cylinder deactivation.

_____ 5. A type of hybrid transmission that often uses two electric motors in combination with an ICE. The two electric motors and the ICE transfer power through a planetary gear set, allowing an infinite amount of gear ratios.

_____ 6. A type of braking in which the kinetic energy of the vehicle's motion is captured rather than being lost to heat in a conventional braking system.

_____ 7. A type of transmission that has no fixed gears, as in a conventional transmission, but rather can adjust gear ratios infinitely within the design of the transmission.

_____ 8. Use of an electric motor to supplement the engine's torque whenever needed, allowing for a smaller ICE to be used.

_____ 9. A Honda hybrid drive system that uses a moderate-sized electric motor installed between the engine and the transmission.

Multiple Choice

Read each item carefully, and then select the best response.

_____ 1. When the power train control module shuts off the hybrid engine at a stoplight, it is an example of _____.
   A. torque smoothing
   B. regenerative braking
   C. torque assist
   D. idle stop

_____ **2.** What ability can a hybrid employ to smooth out the power pulses of the ICE and create a flatter torque output curve?
   A. Torque assist
   B. Torque smoothing
   C. Idle stop
   D. Regenerative braking

_____ **3.** When a conventional vehicle is being stopped, most of the kinetic energy of the vehicle's movement is converted into _____.
   A. heat
   B. electricity
   C. electromagnetic energy
   D. All of the above

_____ **4.** What ability of a hybrid engine reduces the need to use the internal combustion engine at lower rpms by using an electric motor to help propel the vehicle from a stop?
   A. Torque smoothing
   B. Kinetic braking
   C. Torque assist
   D. Integrated propulsion

_____ **5.** What hybrid system is classified as a parallel hybrid because the electric motor is operating at the same time as the gasoline engine?
   A. Two-mode
   B. Integrated motor assist
   C. Belt alternator starter
   D. Continuously variable

_____ **6.** What type of transaxle can be used in combination with the integrated motor assist system?
   A. Automatic transmission
   B. Manual transmission
   C. Continuously variable transmission
   D. All of the above

_____ **7.** The Toyota and Lexus hybrids are classified as _____ because either the internal combustion engine or the electric motor/generator can propel the vehicle, or both can be used together.
   A. series-parallel hybrid
   B. parallel hybrid
   C. series hybrid
   D. integrated hybrid

_____ **8.** A continuously variable transmission has _____ fixed gear ratios.
   A. three
   B. four
   C. five
   D. zero

_____ **9.** All of the following are basic types of continuously variable transmissions commonly used in production vehicles *except*_____.
   A. variable-diameter pulley
   B. integrated motor assist
   C. reeves drive
   D. electronic continuously variable

_____ **10.** The _____ type of continuously variable transmission can be found in both Toyota and Ford hybrid vehicles.
   A. ECVT
   B. toroidal
   C. variable-diameter pulley
   D. reeves drive

True/False

If you believe the statement to be more true than false, write the letter "T" in the space provided. If you believe the statement to be more false than true, write the letter "F".

_____ 1. Automobile manufacturers have had hybrid vehicles and vehicles with continuously variable transmissions in mass production for more than 20 years.

_____ 2. Mechanical brakes can be used to recapture kinetic energy while braking.

_____ 3. Hybrid vehicles are able to maintain a higher gas mileage rating during highway driving than in stop-and-go traffic.

_____ 4. An electric motor is capable of creating its maximum torque as soon as it begins spinning.

_____ 5. On many full hybrids, the vehicle can operate at low speeds using the electric motor only.

_____ 6. Currently, Honda integrated motor assist hybrids cannot drive using the electric motor only.

_____ 7. During electric-only mode, the ICE on Toyota and Lexus hybrids is still used to continually charge the batteries.

_____ 8. Some of the early CVTs in Europe used a stiff rubber belt, as do many snowmobiles and all-terrain vehicles.

_____ 9. As the vehicle gains speed, the input-pulley diameter of the VDP increases, while the output-pulley diameter decreases.

_____ 10. Continuously variable transmissions require special transmission fluid.

Fill in the Blank

Read each item carefully, and then complete the statement by filling in the missing word(s).

1. Many hybrid vehicles that use idle stop have a small, electric transmission _____ _____ used to prevent a delay in the engagement of the transmission when the vehicle is restarted.

2. On a hybrid vehicle, when the driver initiates a stop, the electric motor becomes a _____.

3. The _____ _____ _____ system uses a 42-volt battery.

4. The integrated motor assist system uses a thin electric _____ in place of a conventional flywheel or flex plate.

5. Toyota and Lexus hybrids use voltages from _____ to _____ volts, depending on the application.

6. The only major difference between the Ford hybrid and the Toyota hybrid is that rather than having the two electric motor/generators directly connected to the ring gear and the sun gear, they are attached through a set of _____ _____.

7. A continuously variable transmission has a transmission that can infinitely change the _____ _____ within its operational design.

8. Each of the VDP pulleys has two movable drive faces called _____.

9. The most common type of continuously variable transmission is called a variable-diameter pulley or _____ drive CVT.

10. The _____ _____ is made up of hundreds of transversely mounted steel plates that are held in place with several steel bands running longitudinally around the edge of the plates.

Labeling

Label the following diagrams with the correct terms.

1. Planetary gear operation as the engine is started:

 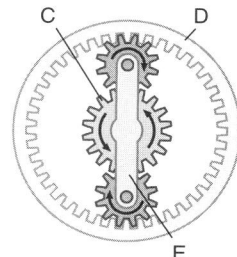

A. _____   Status: _____

B. _____   Status: _____

C. _____   Status: _____

D. _____   Status: _____

E. _____   Status: _____

F. _____   Status: _____

2. Cutaway of a Ford hybrid transmission:

A. _____

B. _____

C. _____

D. _____

E. _____

F. _____

3. A VDP CVT—two sheaves that are moveable:

A. _____

B. _____

C. _____

D. _____

4. The changing sizes of the input and output pulleys:

A. _____

B. _____

C. _____

D. _____

Crossword Puzzle

Use the clues in the column to complete the puzzle.

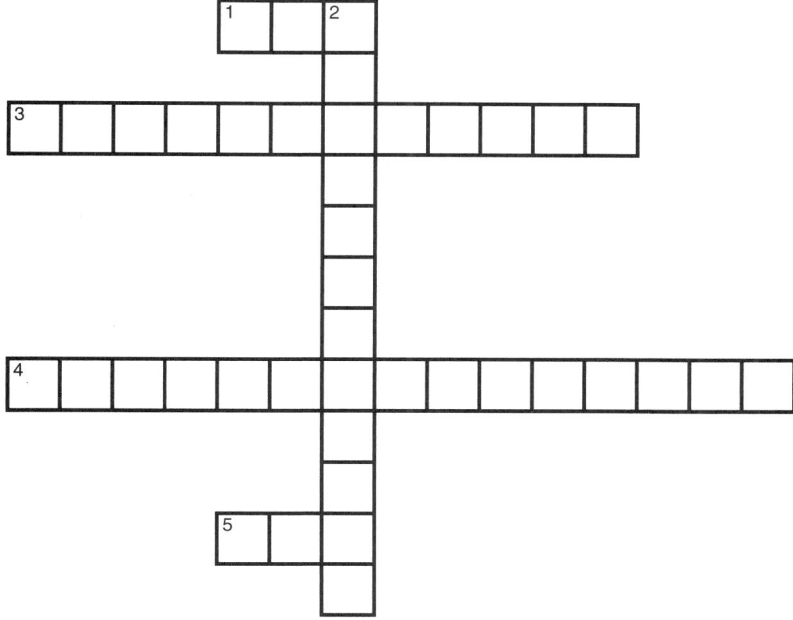

Across

1. A type of transmission that has no fixed gears, as in a conventional transmission, but rather can adjust gear ratios infinitely within the design of the transmission.

3. A type of braking in which the kinetic energy of the vehicle's motion is captured rather than being lost to heat in a conventional braking system.

4. A process that uses an electric motor to smooth out engine power pulses when an ICE is operating at low rpm or when the vehicle is using fuel management techniques such as cylinder deactivation.

5. A type of hybrid drive system that uses a belt-driven alternator/starter that operates on 42 volts.

Down

2. Use of an electric motor to supplement the engine's torque whenever additional torque is needed, allowing for a smaller ICE to be used.

ASE-Type Questions

Read each item carefully, and then select the best response.

_____ **1.** Tech A says that hybrid vehicles have an internal combustion engine and an electric motor. Tech B says that most hybrid vehicles utilize regenerative braking to help improve fuel economy. Who is correct?
  A. Tech A
  B. Tech B
  C. Both A and B
  D. Neither A nor B

_____ **2.** Tech A says that it is critical to have the proper safety equipment before working on a hybrid. Tech B says that hybrid vehicles can start up or move on their own if not powered down properly. Who is correct?
  A. Tech A
  B. Tech B
  C. Both A and B
  D. Neither A nor B

_____ 3. Tech A says that hybrid vehicles will typically use the same transmission as a non-hybrid vehicle. Tech B says that some hybrid transmissions have a small electric fluid pump for when the transmission is in idle/stop mode. Who is correct?

A. Tech A

B. Tech B

C. Both A and B

D. Neither A nor B

_____ 4. Tech A says that BAS vehicles use an alternator to help slow the vehicle and act as a starter. Tech B says that the BAS alternator can be used to propel the vehicle without the ICE. Who is correct?

A. Tech A

B. Tech B

C. Both A and B

D. Neither A nor B

_____ 5. Tech A says that the transmission in a Toyota hybrid has a separate gear to provide reverse. Tech B says that Toyota hybrids use a VDP CVT transmission. Who is correct?

A. Tech A

B. Tech B

C. Both A and B

D. Neither A nor B

_____ 6. Tech A says that some VDP CVTs use a steel belt to transfer power from pulley to pulley. Tech B says that regeneration in a hybrid vehicle is highest during acceleration. Who is correct?

A. Tech A

B. Tech B

C. Both A and B

D. Neither A nor B

_____ 7. Tech A says that some CVTs require a heating system to warm the transmission fluid during cold weather operation. Tech B says that CVTs use conventional automatic transmission fluid. Who is correct?

A. Tech A

B. Tech B

C. Both A and B

D. Neither A nor B

_____ 8. Tech A says that on a VDP CVT, each pulley has a moveable sheave. Tech B says that most CVTs in vehicles use a rubber drive belt between two movable pulleys. Who is correct?

A. Tech A

B. Tech B

C. Both A and B

D. Neither A nor B

_____ 9. Tech A says that regeneration occurs during braking. Tech B says that regeneration charges the main battery pack. Who is correct?

A. Tech A

B. Tech B

C. Both A and B

D. Neither A nor B

_____ 10. Tech A says that the Honda IMA system uses a thin electric motor between the engine and the CVT transmission. Tech B says that the Honda IMA system uses voltages between 144 and 158 volts. Who is correct?

A. Tech A

B. Tech B

C. Both A and B

D. Neither A nor B

Manual Transmission/ Transaxle Principles

© AbleStock

Chapter Review

The following activities have been designed to help you refresh your knowledge of this chapter. Your instructor may require you to complete some or all of these activities as a regular part of your *Fundamentals of Automotive Technology* training program. You are encouraged to complete any activity that your instructor does not assign as a way to enhance your learning in the classroom.

Matching

Match the following terms with the correct description or example.

A. Accelerator pedal
B. Axial load
C. Dead axle
D. Drive axle
E. Drive train
F. Gear set
G. Helical gears

H. Independent rear axle
I. Live axle
J. Power flow
K. Radial load
L. Rotational speed
M. Shaft
N. Solid rear axle

O. Solid axle
P. Splined
Q. Spur gears
R. Thrust washers
S. Transfer case
T. Transmission

_____ 1. Gears that have teeth set on an angle to the gear face; they operate more quietly than spur gears.

_____ 2. An assembly that houses a variety of gear sets that allow the vehicle to be driven at a wider range of speeds and terrain conditions than would be possible without a transmission.

_____ 3. The load that is perpendicular to a shaft, usually controlled by bearings or bushings.

_____ 4. Gears with straight-cut gear teeth.

_____ 5. A type of axle that has a one-piece axle housing, so the action of hitting a bump with one wheel affects the other wheel.

_____ 6. Flat, washer-shaped bearings that provide a wear surface between two rotating components that are loaded axially.

_____ 7. An axle that provides power to a wheel.

_____ 8. Typically, a shaft and gear that have parallel grooves machined in them so they mate with each other and lock together rotationally.

_____ 9. An axle that provides power to the wheels.

_____ 10. The load applied in line with a shaft. It can be controlled with thrust bearings.

_____ 11. An assembly used in four-wheel drive vehicles to transmit power to either two wheels only or all four wheels.

_____ 12. The component assemblies that transmit power from the engine all the way to the drive wheels.

_____ 13. The path that power takes from the beginning of an assembly to the end. In a transmission, power flow changes as different gears are selected by the driver.

_____ 14. The foot-operated pedal used by the driver to increase and decrease the amount of power the engine develops.

_____ 15. The speed at which an object rotates, measured in revolutions per minute (rpm).

_____ 16. An axle that supplies no power to the wheels.

_____ 17. A type of axle that is not flexible, with splines on one end to fit the final drive unit and a flange on the other end to power the wheel.

_____ 18. Two or more gears that are in mesh with each other.

_____ **19.** A type of rear-suspension system that allows each wheel on the axle to move independently of the other.

_____ **20.** The long, narrow component that carries one or more gears or has gears machined into it.

Multiple Choice

Read each item carefully, and then select the best response.

_____ **1.** In 1894, Louis René Panhard and Émile Levassor designed a _____.
  A. rear-wheel drive carriage
  B. synchromesh transmission
  C. automatic transmission
  D. multi-gear manual transmission

_____ **2.** Who connected an engine to a transmission and created a live rear axle by using a metal axle shaft supported by bushings in 1898?
  A. C. E. Duryea
  B. Louis Renault
  C. Henry Ford
  D. Ferdinand Porsche

_____ **3.** In 1928, Cadillac introduced the first _____.
  A. synchromesh transmission
  B. front-wheel drive automobile
  C. planetary gear set
  D. multi-gear transmission

_____ **4.** The amplification of the input force by trading distance moved for greater output force is the definition of _____.
  A. ratio
  B. mechanical advantage
  C. power flow
  D. rotational speed

_____ **5.** Given a 4:1 mechanical advantage, if a person pushes the lever down with a force of 100 pounds, the force the lever generates against the rock is _____.
  A. 140 lb
  B. 40 lb
  C. 400 lb
  D. 4 lb

_____ **6.** If the drive gear has 15 teeth and the driven gear has 30 teeth, then the gear ratio is _____.
  A. 2:1
  B. 800 lb
  C. 1:4
  D. 4 lb

_____ **7.** The path in which power is transmitted through a series of components is called _____.
  A. linkage
  B. power flow
  C. drive train
  D. transmission

_____ **8.** What term relates to layouts where the transmission and final drive are integrated into a common assembly and is usually used on front-wheel drive vehicles?
  A. Differential
  B. Drive train
  C. Torque converter
  D. Transaxle

_____ **9.** What transmits power from the transmission to the final drive assembly on rear-wheel drive vehicles?
  A. Clutch plate
  B. Transaxle
  C. Drive shaft
  D. Live axle

_____ **10.** Flexible drive axles are called _____.
- **A.** radial axles
- **B.** dead axles
- **C.** half-shafts
- **D.** shafts

_____ **11.** What provides front-wheel drive axles the ability to change length as the vehicle goes over bumps and dips and allows for the wheels to be steered?
- **A.** Constant velocity joints
- **B.** Universal joints
- **C.** Drive shafts
- **D.** Synchronizers

_____ **12.** What component(s) allow the axles to turn at different speeds when the vehicle is cornering or turning?
- **A.** Constant velocity joints
- **B.** Universal joints
- **C.** Differential assembly
- **D.** Transfer case

_____ **13.** If one wheel is stuck in the snow or ice a(n) _____ allows both of the rear wheels to supply power to the ground in order to continue forward motion.
- **A.** transfer case
- **B.** limited slip differential
- **C.** transaxle
- **D.** open differential

_____ **14.** What term refers to grease's ability to be a solid but while under stress to flow or become thin in order to lubricate properly?
- **A.** Thixotropy
- **B.** Viscosity
- **C.** Viscidity
- **D.** Glutinousness

_____ **15.** Fluid leaks may be a result of _____.
- **A.** too much fluid in the transmission
- **B.** improper fluid type
- **C.** transmission case porosity or cracks
- **D.** All of the above

True/False

If you believe the statement to be more true than false, write the letter "T" in the space provided. If you believe the statement to be more false than true, write the letter "F".

_____ **1.** Prior to 1898, vehicles were either belt or chain driven.

_____ **2.** If the output force is four times greater than the input force, the input distance moved is two times greater than the output distance.

_____ **3.** As the gear ratio decreases, the output speed increases.

_____ **4.** The clutch pedal is used for acceleration or deceleration of the engine.

_____ **5.** The final drive assembly incorporates a set of differential gears arranged so they sit between the two axles.

_____ **6.** Shafts are used to support gears and are machined precisely to accommodate bearings and individual gears.

_____ **7.** Splines allow shafts and gears to have greater rotational speeds, while at the same time minimizing metal-to-metal friction, which contributes to increased fuel economy and longer transmission life.

_____ **8.** The only difference between the transmission and the transaxle is that the transaxle incorporates the final drive assembly in its construction.

_____ **9.** Transmissions/transaxles are rated by the manufacturers for how much twisting force, measured in foot-pounds, they can handle.

_____ **10.** All manual transmissions use the same type of fluid.

Fill in the Blank

Read each item carefully, and then complete the statement by filling in the missing word(s).

1. A(n) _____ _____ applies a friction device between the gear and the shaft to match the gear speed to the shaft speed.

2. A gear set in which the _____ (drive) gear has half as many teeth as the _____ (driven) gear has a gear ratio of 2:1.

3. Driving a large gear with a smaller gear results in a gear _____.

4. The _____ system is the medium by which the driver can connect and disconnect the engine from the transmission, resulting in the vehicle's forward or rearward movement.

5. The final _____ _____ gives the final gear reduction to the drive train and powers the drive wheels through axles.

6. The drive shaft uses _____ _____, which allow the drive shaft to change angles due to the movement of the suspension relative to the body.

7. Both front-wheel drive and rear-wheel drive vehicles use _____ _____ to power the wheels.

8. Depressing the clutch pedal closes the contacts of the _____ _____ switch.

9. _____ are round parts with teeth cut on the outside perimeter.

10. In a manual transmission vehicle, the component that locks the engine and transmission together is the _____ system.

11. In a four-wheel drive vehicle, the _____ _____ takes the power from the transmission and directs it to one or both axles, depending on the mode selected.

12. With the _____ _____ assembly, if one wheel is stuck in the snow or ice, the other wheel cannot supply power.

13. The _____ _____ drive axle uses one half-shaft axle for each of the two wheels.

14. A(n) _____ _____ allows the wheels to freely rotate on the axle assembly and to not drive the wheels.

15. _____ maintenance is designed to extend the service life of the transmission, improve the reliability of the vehicle, and help prevent vehicle breakdowns that could leave the driver and passengers stranded.

Labeling

Label the following diagrams with the correct terms.

1. Power flow in a transmission:

A. _____

B. _____

C. _____

D. _____

2. Typical final drive assembly:

A. _____

B. _____

C. _____

3. Typical drive shaft:

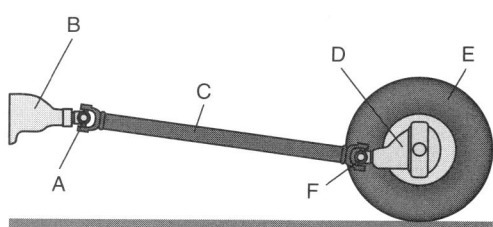

A. _____

B. _____

C. _____

D. _____

E. _____

F. _____

4. Axial-loaded thrust washer:

A. _____

B. _____

5. Clutch components designed to connect and disconnect power to the transmission:

A. _____

B. _____

C. _____

D. _____

E. _____

F. _____

Skill Drills

Test your knowledge of skill drills by filling in the correct words in the photo captions.

1. Checking the Fluid Level of a Manual Transmission:

Step 1: Raise the vehicle on the lift. Inspect the _____ for leaks. Remove the filler plug using the proper _____. Inspect the filler plug and fill hole for _____ _____, and replace/repair if necessary.

Step 2: If the gearbox fluid begins to _____ _____ as the filler plug is removed, let the gearbox fluid seek its own level before _____ the filler plug. The gearbox fluid level should be at the _____ of the filler plug hole.

Step 3: If the fluid level is low, refill with the specified fluid, reinstall the filler plug, and wipe the area around the filler plug hole with a clean _____ _____. Tighten the filler plug to the specified _____.

2. Checking and Adjusting the Differential/Transfer Case Fluid Level:

Step 1: Raise the vehicle on an approved lift. Inspect the _____ and _____ _____ before checking the fluid level. Obtain a clean _____ _____ before removal of the filler plug, as fluid may spill out. Remove the filler plug using the proper wrench. Inspect the filler plug for _____ _____, and replace if necessary. Inspect the threads in the differential and transfer case fill holes for damage also.

Step 2: If the fluid begins to run out as the filler plug is removed, let the fluid seek its own _____ before reinstalling the filler plug. The fluid level should be at the bottom of the _____ _____ _____.

Step 3: If the _____ _____ is low, refill with the specified fluid, reinstall the filler plug, and _____ the area around the filler plug hole with a clean shop towel. _____ the filler plug to the specified torque.

3. Changing the Gearbox Fluid:

Step 1: _____ the vehicle using an approved lift. Obtain a clean drain pan to put the used _____ in.

Step 2: Inspect the _____ for leaks.

Step 3: Remove the drain plug from the _____ of the transmission, being careful of the _____ gearbox fluid. Let gearbox fluid drain until it has stopped running. If necessary, drain the _____ _____ assembly in the same manner.

Step 4: Replace the _____ _____(s) and tighten to specification, and _____ the fill plug(s).

Step 5: Refill the transmission and final drive to the proper level using _____-approved gearbox fluid. _____ the fill plug(s) and tighten. Use a shop towel to wipe away any spillage. _____ _____ the vehicle. If necessary, put the vehicle back on the lift and check for any leaks that may have resulted from the service.

4. Identifying the Cause of Fluid Loss in a Transmission:

Step 1: Raise the vehicle on a _____ or jack stands. Look for leaks in the transmission _____ _____ to the engine block (front seal).

Step 2: Look for leaks in the transmission _____ outlet. Look for leaks in all case _____ areas.

Step 3: Look for leaks in the _____ tail shaft _____.

Step 4: Inspect for transmission case _____ such as _____ and _____.

Step 5: Look for _____ at the drain and fill plugs.

Step 6: Check the gear fluid level as you are removing the drain plug, and look for evidence of excess fluid. If found, let the excess fluid drain into a container. If necessary, use your _____ _____ to check the level. If the fluid is more than half an inch below the bottom of the threaded hole, add new fluid of the correct _____ and type.

Crossword Puzzle

Use the clues in the column to complete the puzzle.

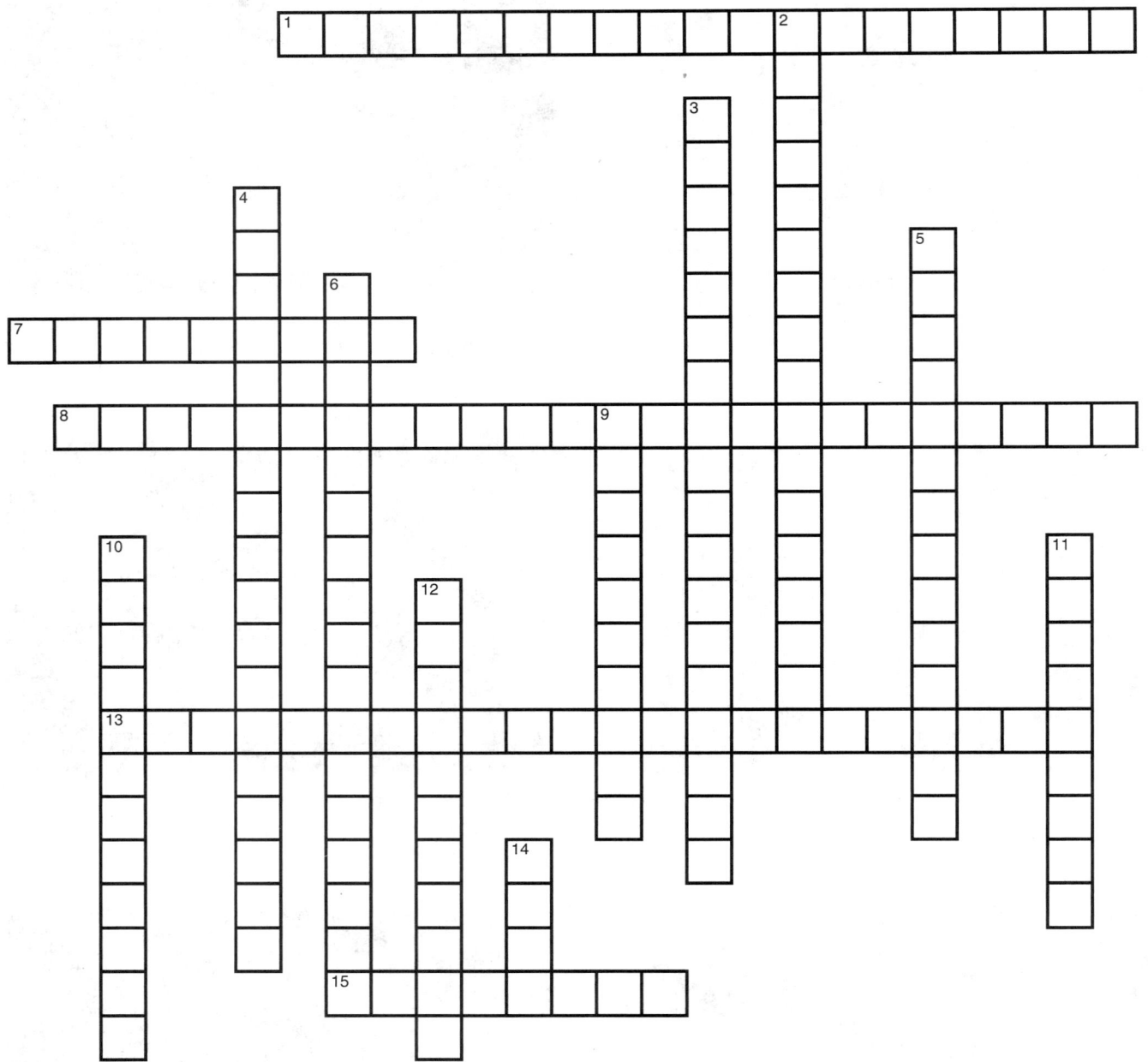

Across

1. The process of using a device to get more output force than the amount of input force, with the trade-off being that the input distance is proportionately longer than the output distance.

7. The description of the difference in speed between gears in mesh, determined by comparing the number of teeth on each gear (drive/driven).

8. A differential assembly that allows both axles to turn at their own speed when turning a corner, but is dependent on the traction of the tires to deliver torque to the ground. If one wheel has no traction, all of the engine's torque will be used at that wheel, causing it to simply spin.

13. Joints commonly used in front-wheel drive vehicles to allow flexibility of the axle while turning.

15. The spring-steel C-shaped ring that is fitted in a groove and holds gears, bearings, and shafts in place.

Down

2. The components that make up the drive axle including the axles, final drive assembly, bearings, and axle housing.

3. An electrical switch that is operated by the clutch pedal and keeps the starter motor from cranking the engine over until the clutch is fully depressed.

4. An assembly used to power the drive wheels and allow the wheels to rotate at different speeds as the vehicle turns.

5. A cross-shaped joint with bearings on each leg where one set of parallel legs is connected to the end of one shaft and the other set of parallel legs is connected to the end of a second shaft. This arrangement allows the shafts to operate at shallow angles to each other.

6. Gears situated in the final drive assembly that are meshed together and with both axles, allowing the wheels to rotate at different speeds when turning a corner.

9. The ability of a semisolid grease to flow when agitated or stressed.

10. A mechanically operated assembly that connects and disconnects the engine from the transmission.

11. An axle that has CV joints on each end and that fits between the transaxle and wheel. Typically, one is used on each side of a vehicle.

12. The foot-operated pedal used by the driver to engage and disengage the clutch.

14. A relatively round, rotating part with internal or external teeth that are designed to mesh with another gear for the purpose of transmitting torque.

ASE-Type Questions

Read each item carefully, and then select the best response.

_____ **1.** Tech A says that friction bearings are made up of balls and rollers. Tech B says that non-friction bearings are in sliding contact between moving surfaces. Who is correct?
  A. Tech A
  B. Tech B
  C. Both A and B
  D. Neither A nor B

_____ **2.** Tech A says that gear ratios are all the same in any five-speed transmission. Tech B says that gear ratios vary from transmission to transmission. Who is correct?
  A. Tech A
  B. Tech B
  C. Both A and B
  D. Neither A nor B

_____ **3.** Tech A says that a road test is helpful in diagnosing transmission issues. Tech B says that having a thorough understanding of the customer concern is important when diagnosing transmission issues. Who is correct?
  A. Tech A
  B. Tech B
  C. Both A and B
  D. Neither A nor B

_____ **4.** Tech A says that gear lube can be used in all manual transmissions. Tech B says that some manual transmissions use engine oil as a lubricant. Who is correct?
  A. Tech A
  B. Tech B
  C. Both A and B
  D. Neither A nor B

_____ **5.** Tech A says that the final drive is used to provide an increase in the rotational speed of the axles. Tech B says that the final drive is used to provide an increase of twisting force to the axles. Who is correct?
  A. Tech A
  B. Tech B
  C. Both A and B
  D. Neither A nor B

_____ 6. Tech A says that a gear set that has a drive gear with 9 teeth and a driven gear with 27 teeth has a gear ratio of 3:1. Tech B says that the drive gear is also called the output gear. Who is correct?
   A. Tech A
   B. Tech B
   C. Both A and B
   D. Neither A nor B

_____ 7. Tech A says that transmission fluid levels are critical to the life of the transmission. Tech B says that transmission oil should be changed when changing the engine oil. Who is correct?
   A. Tech A
   B. Tech B
   C. Both A and B
   D. Neither A nor B

_____ 8. Tech A says that a road test is part of a good process for diagnosing a customer drive train complaint. Tech B says that the ring gear and pinion gear are part of the final drive assembly. Who is correct?
   A. Tech A
   B. Tech B
   C. Both A and B
   D. Neither A nor B

_____ 9. Tech A says that the clutch uses a cone-style synchronizer to match the speed of the engine to the manual transmission. Tech B says that the fill plug hole is where to check the fluid level in most manual transmissions. Who is correct?
   A. Tech A
   B. Tech B
   C. Both A and B
   D. Neither A nor B

_____ 10. Tech A says that the differential assembly provides a means for the inside and outside wheels to turn at different speeds when going around a corner. Tech B says that the differential assembly provides smooth shifts and reduces gear "grinding" by matching gear speeds. Who is correct?
   A. Tech A
   B. Tech B
   C. Both A and B
   D. Neither A nor B

The Clutch System

Chapter Review

The following activities have been designed to help you refresh your knowledge of this chapter. Your instructor may require you to complete some or all of these activities as a regular part of your *Fundamentals of Automotive Technology* training program. You are encouraged to complete any activity that your instructor does not assign as a way to enhance your learning.

Matching

Match the following terms with the correct description or example.

A. Carrier	**H.** Free-play	**O.** Release mechanisms
B. Clutch disc	**I.** Friction facing	**P.** Single-plate clutch
C. Clutch fork	**J.** Fulcrum ring	**Q.** Slave cylinder
D. Coefficient of friction	**K.** Pilot bearing	**R.** Throw-out bearing
E. Coil-spring pressure plate	**L.** Push-type clutch	**S.** Torsional vibrations
F. Diaphragm pressure plate	**M.** Pressure plate	**T.** Transmission input shaft
G. Flywheel	**N.** Quadrant ratchet	

_____ **1.** The bearing or bushing that supports the front of the transmission input shaft.

_____ **2.** A heavy metal disc bolted to the crankshaft that is used to smooth out the engine's power pulses and keep the engine moving through the non-power strokes.

_____ **3.** A type of pressure plate that uses coil springs to provide the clamping force.

_____ **4.** The part of the clutch release mechanism that imparts clutch pedal force to the rotating pressure plate levers.

_____ **5.** The material riveted to each side of the clutch disc that mates to the flywheel and pressure plate. Used to provide friction and a wear surface for the clutch assembly.

_____ **6.** The speeding up and slowing down of a shaft, which happen at a relatively high frequency.

_____ **7.** The device used in some cable-operated clutches to provide self-adjustment as the clutch disc wears.

_____ **8.** The component in a hydraulically operated clutch that converts hydraulic pressure to mechanical movement at the clutch fork.

_____ **9.** The center component of the clutch assembly, with friction material riveted on each side. Also called a clutch plate or friction disc.

_____ **10.** Components that operate the clutch. Usually included are the throw-out bearing and the clutch fork. Some manufacturers include the operating system.

_____ **11.** A slightly conical, spring steel plate used to provide the clamping force for the clutch assembly.

_____ **12.** The part of the throw-out bearing assembly that holds the bearing.

_____ **13.** A clutch assembly that uses only one plate to transfer torque from the engine to the transmission. This is the most common type of light vehicle clutch.

_____ **14.** The part of the clutch linkage that operates the throw-out bearing.

_____ **15.** The amount of clearance in the clutch release mechanism as measured at the clutch pedal.

_____ **16.** The shaft that brings engine torque into the transmission.

_____ **17.** A steel ring that is used as a pivot point for the diaphragm spring in the pressure plate.

_____ **18.** The amount of resistance to movement between any two surfaces that are in contact with each other.

_____ 19. The assembly that applies and removes the clamping force on the clutch disc.

_____ 20. A typical clutch system used in modern vehicles where the clutch fork pushes the release bearing forward to release the friction facing from the pressure plate.

Multiple Choice

Read each item carefully, and then select the best response.

_____ 1. The amount of torque a clutch can transmit is dependent upon which of the following?
   A. Diameter of the clutch
   B. Coefficient of friction
   C. Total spring force
   D. All of the above

_____ 2. The _____ is able to connect to and disconnect from the engine's flywheel through the operation of the clutch.
   A. crankshaft
   B. input shaft
   C. pilot bearing
   D. output shaft

_____ 3. The main purpose of the _____ is to smooth out the power pulses from the pistons during the power strokes.
   A. pressure plate
   B. throw-out bearing
   C. flywheel
   D. clutch disc

_____ 4. Which of the following is a type of flywheel?
   A. One-piece
   B. Harmonic
   C. Dual mass
   D. Both A and C

_____ 5. Refinishing the flywheel moves the pressure plate toward the engine and away from the throw-out bearing, increasing _____.
   A. free-play
   B. clutch chatter
   C. runout
   D. fuel economy

_____ 6. The snout of the input shaft rides on the _____.
   A. throw-out bearing
   B. hub
   C. pilot bearing
   D. flywheel

_____ 7. Which of the following is a type of clutch operating mechanism?
   A. Linkage style
   B. Hydraulic
   C. Cable style
   D. All of the above

_____ 8. The pilot bearing is the front bearing, and at the other end of the input shaft is the _____ bearing.
   A. passenger
   B. transmission input
   C. clutch release
   D. transmission output

_____ 9. If the clutch pedal has too little _____, the throw-out bearing could remain in contact with the pressure plate levers.
   A. end-play
   B. spring pressure
   C. free-play
   D. diaphragm pressure

_____ 10. The _____ method is used to properly bleed the hydraulic clutch system.
   A. gravity bleeding
   B. manual bleeding
   C. pressure bleeding
   D. any of the above

True/False

If you believe the statement to be more true than false, write the letter "T" in the space provided. If you believe the statement to be more false than true, write the letter "F".

_____ 1. Automotive manual transmission clutches are wet clutches.

_____ 2. Both output shaft torque and speed can be increased at the same time.

_____ 3. The output shaft is connected directly to the wheels through the drive train components and cannot be disconnected.

_____ 4. The function of the dual mass flywheel is to absorb torsional crankshaft vibrations.

_____ 5. Coil spring pressure plate clutches require less pedal effort to operate.

_____ 6. The clutch throw-out bearing and clutch fork work together to compress the pressure plate springs when the clutch pedal is pressed.

_____ 7. Some vehicles do not require a pilot bearing.

_____ 8. In a cable-style clutch control system the slave cylinder is located directly behind the throw-out bearing and pushes directly on it.

_____ 9. When initially placing the transmission into gear, the transmission input shaft must be disconnected from the engine to prevent gear grinding.

_____ 10. Old fluid in the hydraulic system can cause master cylinder and slave cylinder damage.

Fill in the Blank

Read each item carefully, and then complete the statement by filling in the missing word(s).

1. The _____ allows the driver to engage and disengage the engine from the transmission while operating the vehicle.

2. Two or more clutch plates can be used to form a _____ clutch, increasing the number of facings and the torque capacity.

3. The _____ _____ bolts onto the flywheel.

4. The flywheel _____ _____ enables the starter motor drive gear to crank the engine over.

5. The _____ _____ flywheel improves the engine's fuel economy by smoothing out the power pulses and focusing them in the direction of engine rotation.

6. The diaphragm pressure plate is located inside the clutch cover on two _____ _____, held in place by a number of rivets passing through the diaphragm.

7. The clutch disc is also called a _____ _____ _____ or a friction disc.

8. The throw-out bearing carrier slides on the sleeve of the front _____ _____ that extends from the front of the transmission.

9. The clutch fork _____ is generally screwed into the bell housing and is usually replaceable.

10. While the vehicle is in motion, the transmission of engine _____ must be interrupted for shifting of the gears.

11. Some vehicles use a quadrant _____, which automatically adjusts the clutch pedal free-play as needed when the pedal is lifted by the driver's toe.

12. In hydraulic clutch release mechanisms, the clutch pedal acts on a master cylinder connected by a hydraulic tube and flexible hose to a _____ cylinder mounted on (external type) or in (internal type) the transmission bell housing.

13. Older technology clutch operating systems used a series of levers with an equalizing mechanism called a(n) _____ _____.

14. As the clutch wears, the _____ _____ becomes thinner.

15. In the case of a hydraulic clutch system failure, it may be necessary to _____ the air from the system.

Labeling

Label the following diagrams with the correct terms.

1. Standard light vehicle clutch:

A. _____

B. _____

C. _____

D. _____

E. _____

F. _____

G. _____

H. _____

I. _____

J. _____

K. _____

L. _____

M. _____

N. _____

O. _____

P. _____

Q. _____

2. Releasing the clutch pedal:

Driven unit disengaged

A. _____

B. _____

C. _____

D. _____

E. _____

F. _____

G. _____

H. _____

Driven unit engaged

3. Two types of dual mass flywheels:

A

A. _____

B. _____

C. _____

D. _____

E. _____

F. _____

G. _____

H. _____

I. _____

J. _____

K. _____

4. Diaphragm pressure plate:

A. _____

B. _____

C. _____

D. _____

E. _____

F. _____

5. Coil spring pressure plate:

A. _____

B. _____

C. _____

D. _____

E. _____

6. Clutch disc components:

A. _____

B. _____

C. _____

D. _____

E. _____

F. _____

G. _____

7. Waved springs:

A. _____

B. _____

C. _____

D. _____

E. _____

8. Cable-operated clutch:

A. _____

B. _____

C. _____

D. _____

E. _____

F. _____

G. _____

H. _____

I. _____

J. _____

9. Hydraulic clutch control:

A. _____

B. _____

C. _____

D. _____

E. _____

F. _____

G. _____

H. _____

I. _____

10. Lever-operated system:

A. _____

B. _____

C. _____

D. _____

E. _____

F. _____

G. _____

H. _____

I. _____

Crossword Puzzle

Use the clues to complete the puzzle.

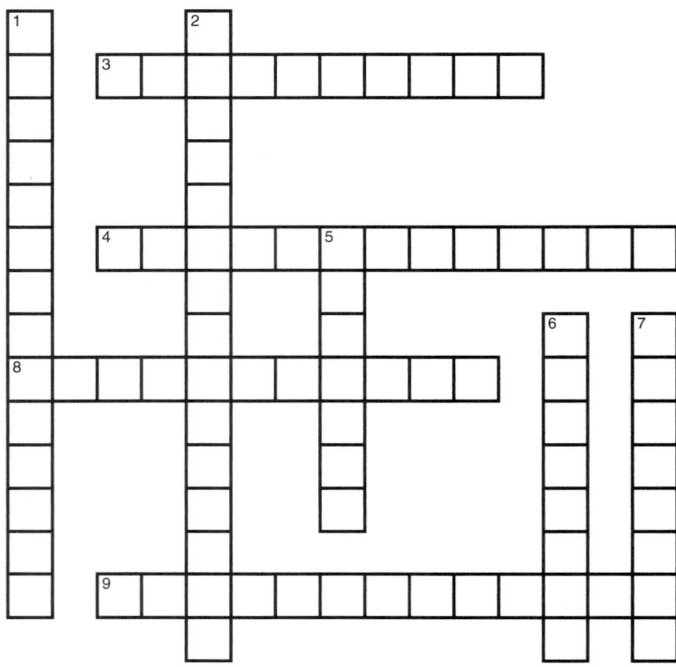

Across

3. The part of the clutch linkage that operates the throw-out bearing.

4. The component in a hydraulically operated clutch that converts hydraulic pressure to mechanical movement at the clutch fork.

8. A steel ring that is used as a pivot point for the diaphragm spring in the pressure plate.

9. The assembly that applies and removed the clamping force on the clutch disc.

Down

1. The material riveted to each side of the clutch disc that mates to the flywheel and pressure plate. Used to provide friction and a wear surface for the clutch assembly.

2. The device used in some cable-operated clutches to provide self-adjustment as the clutch disc wears.

5. The part of the throw-out bearing assembly that holds the bearing.

6. The amount of clearance in the clutch release mechanism as measured at the clutch petal pedal.

7. A heavy metal disc bolted to the crankshaft that is used to smooth out the engine's power pulses and keep the engine moving through the non-power strokes.

ASE-Type Questions

Read each item carefully, and then select the best response.

_____ **1.** Tech A says that the pressure plate friction surface rides on the flywheel friction surface to transmit torque. Tech B says that the flywheel can either be flat or stepped. Who is correct?
   A. Tech A
   B. Tech B
   C. Both A and B
   D. Neither A nor B

_____ **2.** Tech A says that insufficient clutch pedal clearance (free-play) can cause gear clashing when shifting. Tech B says that when the engine is idling and the clutch pedal is released, the friction disc should stop rotating. Who is correct?

   A. Tech A
   B. Tech B
   C. Both A and B
   D. Neither A nor B

_____ **3.** Tech A says a dual mass flywheel improves the engine's fuel economy by smoothing out the power pulses and focusing them in the direction of engine rotation. Tech B says there are two types of dual mass flywheel. Who is correct?

   A. Tech A
   B. Tech B
   C. Both A and B
   D. Neither A nor B

_____ **4.** Tech A says that on larger vehicles, such as trucks, a ball bearing type may be used for a pilot bearing. Tech B says that in some applications, pilot bearings may be placed or pressed into the center of the flywheel. Who is correct?

   A. Tech A
   B. Tech B
   C. Both A and B
   D. Neither A nor B

_____ **5.** Tech A says that there are three main types of mechanical systems used for clutch operation. Tech B says that a hydraulically operated system is also used in some applications. Who is correct?

   A. Tech A
   B. Tech B
   C. Both A and B
   D. Neither A nor B

_____ **6.** Tech A says that it is important to check the clutch linkage mechanism for proper operation and correct the adjustment (free-play) periodically. Tech B says that if the clutch pedal has too little free-play, the throw-out bearing could remain in contact with the pressure plate levers. Who is correct?

   A. Tech A
   B. Tech B
   C. Both A and B
   D. Neither A nor B

_____ **7.** Tech A says that the pilot bearing can be a needle-style bearing. Tech B says that the pilot bearing can be a brass bushing style. Who is correct?

   A. Tech A
   B. Tech B
   C. Both A and B
   D. Neither A nor B

_____ **8.** Tech A says that when bleeding a clutch hydraulic system, one method uses pressure or vacuum to push or pull fluid and air from the system. Tech B says that this method does not require special bleeding tools or equipment. Who is correct?

   A. Tech A
   B. Tech B
   C. Both A and B
   D. Neither A nor B

_____ **9.** Tech A says that in a mechanical system, if the clutch pedal adjustment has too much free-play, there may not have enough travel to fully release the pressure plate. Tech B says that this condition can cause the gears to clash (grind) when shifting and result in heavy synchronizer wear. Who is correct?

   A. Tech A
   B. Tech B
   C. Both A and B
   D. Neither A nor B

_____ **10.** Two technicians are discussing pressure plates. Tech A says that diaphragm pressure plates require more pedal effort than coil spring pressure plates. Tech B says that the two types require different free pedal adjustment procedures. Who is correct?

   A. Tech A
   B. Tech B
   C. Both A and B
   D. Neither A nor B

Basic Drive Layouts

Chapter Review

The following activities have been designed to help you refresh your knowledge of this chapter. Your instructor may require you to complete some or all of these activities as a regular part of your Fundamentals of Automotive Technology training program. You are encouraged to complete any activity that your instructor does not assign as a way to enhance your learning.

Matching

Match the following terms with the correct description or example.

A. Backlash clearance
B. Brinelling
C. Companion flange
D. Constant-velocity (CV) joint
E. Dead axle
F. Full floating axle
G. Hypoid bevel gear

H. Live axle
I. Locking hub
J. Longitudinal
K. Power take-off (PTO)
L. Rzeppa joint
M. Side gear
N. Speedy sleeve

O. Torque steer
P. Transfer case
Q. Transverse
R. Tulip/tripod joint
S. Universal joint (U-joint)
T. Viscous coupling

_____ **1.** An aftermarket repair kit that consists of a thin metal sleeve that fits tightly over the seal surface of the axle, providing a new, undamaged surface for the seal to ride against.

_____ **2.** Four-wheel drive front axle hubs that are manually locked or unlocked by turning the knob on the hub.

_____ **3.** A constant velocity joint that has three equally spaced fingers shaped like a star. This configuration allows in-and-out movement of the shaft while allowing flexing.

_____ **4.** A device attached to the transmission that is gear driven and can be used to run accessories such as winches and towing equipment. It can also refer to the gears that send power to the rear axle in a predominantly front-wheel drive vehicle.

_____ **5.** An axle that does not support any weight; if removed, the vehicle will still roll on its wheels.

_____ **6.** A silicone clutch assembly used in all-wheel drive differentials to provide a slight amount of differential action for control of axle rotational speeds.

_____ **7.** The orientation of the engine in which the front of the engine is facing the side of the vehicle.

_____ **8.** Damage done to the surface of a bearing caused by excessive load, which exceeds the limit of the bearing material, typically from shock loads.

_____ **9.** A cross-shaped flexible joint on which caps fit over the ends of the cross. Needle bearings fit between the ends of the cross and the caps, allowing the caps to rotate smoothly.

_____ **10.** A special design of spiral bevel gear, with the centerline of the pinion below the centerline of the ring gear.

_____ **11.** An axle that does not have the capability to drive the vehicle. It is usually found on the rear of front-wheel drive vehicles.

_____ **12.** The orientation of the engine in which the front of the engine is facing the front of the vehicle. It is most commonly found in rear-wheel drive vehicles.

_____ **13.** A type of fixed constant velocity joint that has an inner race, six steel ball bearings, a bearing cage, and an outer race.

_____ **14.** A splined flange that transmits power from the drive shaft to the pinion gear.

_____ **15.** A gear that is splined to the axle shaft and meshes with the spider gears and allows the axles to rotate at their own speeds when cornering and turning.

_____ **16.** The amount of movement between the pinion teeth versus the ring teeth.

_____ **17.** An axle that is powered and can move the vehicle. It is usually found on the rear of rear-wheel drive vehicles.

_____ **18.** A condition in which the vehicle pulls to one side during hard acceleration.

_____ **19.** A joint used to transmit torque through wider angles and without the change of velocity that occurs in U-joints.

_____ **20.** A component that is bolted to the back of the transmission and connects the front and rear axles via the driveshafts.

Multiple Choice

Read each item carefully, and then select the best response.

_____ **1.** The _____ slides in and out on the output shaft, allowing the length changes needed as the suspension moves up and down.
  A. slip yoke
  B. universal joint
  C. half-shaft
  D. speedy sleeve

_____ **2.** A two-piece driveshaft, joined in the middle with splines that can slide on itself to increase and decrease in length is called a _____.
  A. half-shaft
  B. full floating driveshaft
  C. sliding spline driveshaft
  D. Rzeppa joint

_____ **3.** The _____ is a device that transfers torque from one component to another, such as from the transmission output shaft to the final drive assembly.
  A. half-shaft
  B. driveshaft
  C. drive axle
  D. constant velocity joint

_____ **4.** The _____ does not provide any drive capabilities and is used on the rear of front-wheel drive vehicles, and the front of rear-wheel drive vehicles.
  A. dead axle
  B. half-shaft
  C. drive axle
  D. driveshaft

_____ **5.** What type of rear-wheel drive solid axle has a single roller bearing between the hub and the outside of the axle housing?
  A. Dead axle
  B. Semi-floating axle
  C. Three-quarter floating axle
  D. Full-floating axle

_____ **6.** A(n) _____ can be of the single-lip or double-lip design depending on the application.
  A. flange
  B. yoke
  C. speedy sleeve
  D. axle seal

_____ **7.** The use of different-sized axle shafts can cause a problem known as _____.
  A. misalignment
  B. torque steer
  C. slip differential
  D. angularity

_____ **8.** What type of joint allows the shaft to transmit torque through a change of drive angle?
   A. Universal joint
   B. Ball joint
   C. Constant velocity joint
   D. Both A and C

_____ **9.** What type of constant velocity joint does not slide to allow for shaft lengthening or shortening; it simply allows for angle changes as the suspension moves?
   A. Fixed-type joint
   B. Rzeppa joint
   C. Plunge-type joint
   D. Tulip tripod joint

_____ **10.** The _____ differential is a gear-style limited slip differential design that multiplies the torque available from the wheel that is losing traction and turns it over to the slower turning wheel with better traction.
   A. clutch pack
   B. Rzeppa
   C. torsen style
   D. viscous coupling

True/False

If you believe the statement to be more true than false, write the letter "T" in the space provided. If you believe the statement to be more false than true, write the letter "F".

_____ **1.** Longitudinally mounted means that the front of the engine is facing the side of the vehicle.

_____ **2.** In some all-wheel drive vehicles, only two of the wheels are normally powered, typically the front wheels.

_____ **3.** The purpose of the drive axle is to transfer the torque that comes from the engine, transmission, driveshaft, and differential to the wheels and tires, propelling the vehicle forward or backward.

_____ **4.** If a vehicle with a semi-floating axle were to hit a curb and break or snap the axle shaft, the wheel would come off of the vehicle.

_____ **5.** Full floating axles can be removed without removing the wheels.

_____ **6.** A double Cardan joint greatly increases the change in velocity compared to a single Cardan joint.

_____ **7.** Universal joints are larger than constant velocity joints.

_____ **8.** If you hold the ring gear and turn the pinion, you will have a slight clearance back and forth.

_____ **9.** In part-time four-wheel drive vehicles, differentials are fitted to both the front and rear axle assemblies.

_____ **10.** A transfer case is used on vehicles that are primarily front-wheel drive; a power take-off is used on vehicles that are primarily rear-wheel drive.

Fill in the Blank

Read each item carefully, and then complete the statement by filling in the missing word(s).

1. _____-_____ or solid rear axle assemblies enclose the final drive gears, differential gears, and axle shafts into one housing.

2. _____ _____ joints allow for smoother transfer of power and allow for the vehicle to turn more tightly without the joint binding.

3. The _____ can be made up of as many as four separate segments.

4. On a(n) _____-_____ axle, the axle shafts are splined to the differential side gears, and the outer bearing is between the outer end of the axle shaft and the inside of the axle housing.

5. Lug studs are pressed into the axle _____ and used to secure the wheel onto the vehicle.

6. In front-wheel drive vehicles and all-wheel drive vehicles, the driveshafts transfer the drive directly from the _____ _____ inside the transaxle to the front wheels.

7. The _____ _____ is a short section of shaft that typically has a bearing pressed onto it, used to make both half-shafts the same length from left to right.

8. The _____ joint consists of a steel cross with four hardened bearing journals, mounted on needle rollers in hardened caps, which locate the cross in the eyes of the yokes.

9. Often, a sliding spline or a _____-_____ joint is used as the inner half-shaft joint to accommodate for changes in shaft length when traveling on different types of terrain.

10. A CV _____ is used to keep the grease inside the joint, and to keep dirt and debris out.

11. The speed reduction gears in the final drive are called the _____ and _____ gears.

12. A(n) _____-_____ limited slip differential responds very quickly to changes in traction. They also do not bind from friction in turns and do not lose their effectiveness since there are no clutches, like a clutch-style unit.

13. Freewheeling hubs or manually _____ hubs can be installed on the front hubs to disconnect the hubs from the axle shafts and prevent rotation of the front axle components.

14. The transfer case is typically bolted to the rear of the _____.

15. A _____ _____ uses two sets of clutch plates that are alternated front to back. One set of plates is splined to the outer housing, and the other set of plates is splined to the inner housing.

Labeling

Label the following diagrams with the correct terms.

1.

1. _____

2.

2. _____

3.

3. _____

4.

A.

A. _____

B.

B. _____

5.

5. _____

Skill Drills

Test your knowledge of skill drills by filling in the correct words in the photo captions.

1. Diagnosing CV Joint Issues:

Step 1: Thoroughly _____ _____ the vehicle to verify the customer complaint. Safely _____ the vehicle on an approved _____, and make sure it is secure.

Step 2: Visually inspect all _____ CV joints.

Step 3: Look for broken or ruptured _____. Look for _____ or dry-rotted boots, as they are subject to all types of weather.

Step 4: Manually move the axle _____ up and down, looking for unnecessary _____ or bad bearings. These conditions can cause a _____ coming from the side that is bad. They will also cause a _____ sound emanating from the side that is affected.

Step 5: Look for any type of axle damage, such as a _____ axle from a recent_____. Axle damage will cause a _____ from the affected shaft as well.

2. Inspecting Half-Shaft Components:

Step 1: Clamp the entire half-shaft into a soft-jawed _____, and make sure it is secure. Remove the retaining _____ from the CV boot. Slide the _____ down the shaft, paying attention to the condition of the boot.

Step 2: Wipe out as much _____ as possible to be able to access the retaining ring from the CV joint itself. If there is a retaining ring present, _____ the retaining ring with the appropriate tool. When the retaining ring is removed, remove the CV joint from the half-shaft, and inspect the _____ on the end of the half-shaft. This also applies to the other end of the half-shaft.

Step 3: Inspect the old joint to gain an accurate _____ of the failure to prevent a reoccurrence of this failure. Reinstall the new CV joint onto the _____ _____ as required. Apply the lubrication grease that comes with the new CV joint. Tighten the _____ _____ to ensure that grease will not be lost. This applies to both sides of the half-shaft. Reinstall the half-shaft following the _____ guidelines.

3. Inspecting Fluid Leakage:

Step 1: Put the vehicle on a lift, and make sure it is secure. Inspect around the brake _____ where the axle seats for any _____.

Step 2: Inspect the _____ _____ and differential _____ for possible seepage.

Step 3: Check and clean the _____ or _____ for any obstructions that may cause a pressure buildup to occur.

4. Removing and Reinstalling Sealed Wheel Bearings Using the Unitized Wheel Bearing Hub Style:

Step 1: Loosen the axle hub nut, if equipped, while the _____ is still on the _____. Remove the wheel and _____ assembly following the specified procedure. Also _____ the ABS connector and/or sensor if mounted to the hub.

Step 2: If the wheel you are working on is a _____ wheel, remove the axle hub nut and _____ the drive axle.

Step 3: Unbolt and remove the hub assembly from the _____ _____. Clean the knuckle assembly and check the hub _____ for nicks, burrs, or other damage.

Step 4: Make sure that the replacement _____ meets the vehicle specifications by comparing it to the old one.

Step 5: Reassemble the _____ assembly and the ABS _____ following the specified procedure.

Step 6: Install the wheel and _____ the lug nuts. Install the drive axle nut if equipped. Use a new _____ _____ if called for by the manufacturer, and torque to specifications.

5. Inspecting, Removing, and Replacing Front-Wheel Drive Front Wheel Bearings:

Step 1: Raise the front _____ off the ground and secure it with _____ _____. Loosen the lug nuts, and remove the tire and wheel assembly. Remove the _____ from the manual locking hub cap and pull the cap from the axle.

Step 2: Inspect the O-ring on the back side of the _____. Replace it if necessary. Remove the _____ _____ from the hub, and pull the retaining _____ out of the hub. Remove the snap ring from the axle.

Step 3: Remove the _____ _____ from the hub.

Step 4: Remove the bearing retaining nut(s) from the hub assembly to expose the _____. Then remove the hub and _____ assembly for inspection.

Step 5: Pry out the _____ and rear bearing for inspection. Clean and inspect the bearings closely for _____ and _____ due to heat concerns. Replace as necessary.

Step 6: Pack and install new bearings and _____ using the proper bearing race driver. Install the new _____ assembly. Note: When replacing the seal, be sure to put some _____ on the spring inside the seal to prevent it from coming off when the seal is driven into the _____.

Step 7: Reinstall the hub and _____ on the axle, and place the outer wheel bearing in the hub. Thread the bearing adjusting _____ on by hand. Set the wheel bearing _____ following the manufacturer's specifications using the proper tool. Install the locking _____ over the adjusting nut.

Step 8: Thread the _____ _____ up against the locking washer and _____ to the specified torque. If necessary, bend the lock _____ up against the locking nut. Install the locking _____. Install the retaining plate and wire clip.

Step 9: Install a new O-ring on the hub _____, and reinstall the hub cover on the _____. Tighten the hub cover screws to the manufacturer's specifications. _____ _____ the vehicle to ensure that it was repaired properly.

Crossword Puzzle

Use the clues to complete the puzzle.

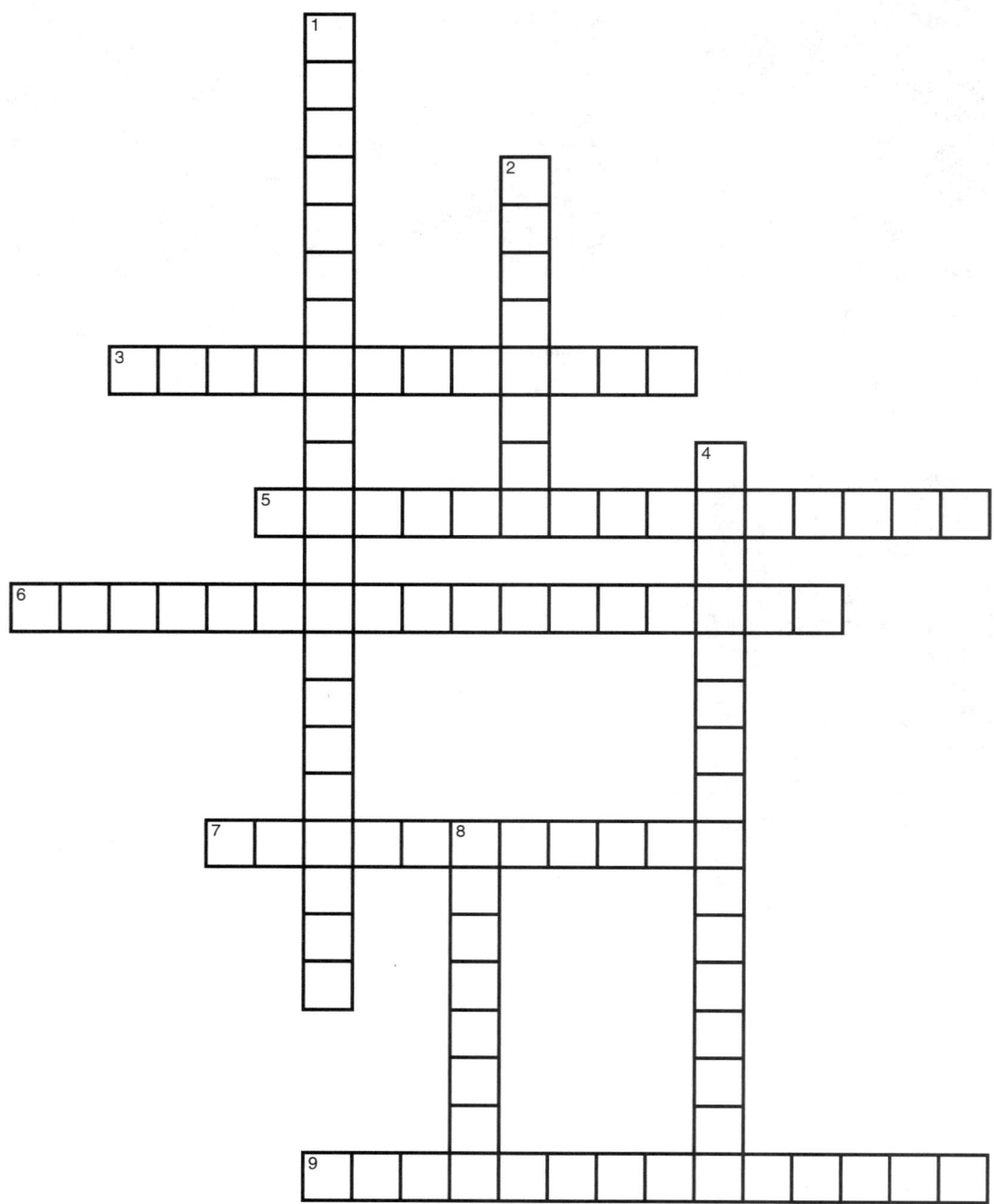

Across

3. A rear-wheel drive axle assembly that has a solid tube incorporating the differential gears.

5. The inner joint on the half-shaft that allows for changes in shaft length.

6. A type of joint that uses two Cardan joints housed in a short carrier and that reduces the change in velocity of a single Cardan joint by using the second joint to cancel out the changes in velocity of the first joint.

7. A joint that consists of a steel cross with four hardened bearing journals, mounted on needle rollers in hardened caps, which locate the cross in the eyes of the yokes. The cross swivels in the yokes as the drive is transferred across the joint.

9. A joint that does not slide to allow for shaft lengthening or shortening; it simply allows for angle changes as the suspension moves.

Down

1. A joint used to transmit torque through wider angles and without the change of velocity that occurs in U-joints.

2. An axle that does not have the capability to drive the vehicle. It is usually found on the rear of front-wheel drive vehicles.

4. An axle that carries the weight of the vehicle; if removed, there is no way to connect the wheel to the vehicle.

8. Part of a two-piece driveshaft that is splined and allows for a change in length of the shaft as the suspension compresses and rebounds.

ASE-Type Questions

Read each item carefully, and then select the best response.

_____ **1.** Tech A says that viscous clutch oil should generally be changed at regular intervals. Tech B says that if a viscous clutch fails, it should be replaced. Who is correct?
A. Tech A
B. Tech B
C. Both A and B
D. Neither A nor B

_____ **2.** Tech A says that all-wheel drive vehicles have a differential in the middle of the drive train. Tech B says that many all-wheel drive vehicles incorporate a viscous clutch in the center differential. Who is correct?
A. Tech A
B. Tech B
C. Both A and B
D. Neither A nor B

_____ **3.** Tech A says that all front wheel hubs lock automatically. Tech B says that all front wheel hubs lock manually. Who is correct?
A. Tech A
B. Tech B
C. Both A and B
D. Neither A nor B

_____ **4.** Tech A says that full floating axles use a single wheel bearing that sits between the axle and axle housing. Tech B says that a differential allows the wheels on a common axle to rotate at different speeds. Who is correct?
A. Tech A
B. Tech B
C. Both A and B
D. Neither A nor B

_____ **5.** Tech A says that part-time four-wheel drive does not use a transfer case. Tech B says that a viscous clutch is located in the rear axle. Who is correct?
A. Tech A
B. Tech B
C. Both A and B
D. Neither A nor B

_____ **6.** Tech A says that constant-velocity joints are used on some four-wheel drive vehicles. Tech B says that a slip joint is part of a driveshaft. Who is correct?
A. Tech A
B. Tech B
C. Both A and B
D. Neither A nor B

_____ **7.** Tech A says that a limited slip differential causes spring-loaded clutch plates to provide torque to both wheels. Tech B says that wrong tire sizes can affect the transfer case operation. Who is correct?

A. Tech A

B. Tech B

C. Both A and B

D. Neither A nor B

_____ **8.** Tech A says that some transfer cases have a chain inside them to transfer torque to the wheels. Tech B says that a rear-wheel drive axle housing can leak due to being overfilled with fluid. Who is correct?

A. Tech A

B. Tech B

C. Both A and B

D. Neither A nor B

_____ **9.** Tech A says that CV boots should be patched if they have cracks or splits. Tech B says that CV joints may use a special type of grease, possibly containing lead. Who is correct?

A. Tech A

B. Tech B

C. Both A and B

D. Neither A nor B

_____ **10.** Tech A says that U-joints can make a chirping or squeaking nose when they are going bad. Tech B says that U-joints are typically used on front wheel drive axles to provide torque through all angles of steering. Who is correct?

A. Tech A

B. Tech B

C. Both A and B

D. Neither A nor B

Servicing Wheels

Chapter Review

The following activities have been designed to help you refresh your knowledge of this chapter. Your instructor may require you to complete some or all of these activities as a regular part of your training program. You are encouraged to complete any activity that your instructor does not assign as a way to enhance your learning.

Matching

Match the following terms with the correct description or example.

A. Bar
B. Casing plies
C. Drop center
D. EH2 rim
E. Negative offset
F. Neutral steer
G. Ply rating
H. Rim flanges
I. Schrader valve
J. Side force

K. Slip angle
L. Steel-disc–type rim
M. Tapered seat
N. Traction grade
O. Tread wear grade
P. Valve core
Q. Valve stem
R. Wheel center
S. Wheel studs
T. Zero offset

_____ **1.** A plain steel wheel that is typically covered by a hubcap.

_____ **2.** A condition in which the plane of the hub mounting surface is positioned toward the brake side or back of the wheel centerline.

_____ **3.** A rubber or steel piece that attaches the tire valve to the rim.

_____ **4.** A one-way valve used in a valve stem.

_____ **5.** A standardized grading system that indicates how well a tire will maintain contact with the road surface when wet.

_____ **6.** A rating system that denotes the number of belt layers or plies that make up the tire carcass. In radial tires, ply rating denotes the relative strength of the plies, not the actual number of plies.

_____ **7.** The part of the wheel containing the holes for the lug studs.

_____ **8.** The specialized rim design that is used with some run-flat tires.

_____ **9.** The number imprinted on the sidewall of a tire by the manufacturer as required by the National Highway Traffic Safety Administration that indicates the tire's tread life.

_____ **10.** A condition in which both the front and the rear tires of a vehicle are experiencing the same slip angle.

_____ **11.** A condition in which the plane of the hub mounting surface is even with the centerline of the wheel.

_____ **12.** The pressure on the wheel that pushes it toward the outside or inside of the rim as the vehicle makes a turn.

_____ **13.** A wheel design with part of the center section of the wheel a smaller diameter than the rest. It is used for mounting and demounting the tire.

_____ **14.** A tire's sideways distortion that makes the vehicle follow a path at an angle to the direction the road wheel is pointing.

_____ **15.** The threaded fasteners that attach the wheel to the vehicle.

_____ **16.** A metric unit of measure for pressure.

_____ **17.** A type of lug nut with a tapered end toward the rim that helps center the wheel on the wheel studs.

_____ **18.** A network of cords that give the tire shape and strength.

_____ **19.** The one-way spring-loaded valve that screws into the valve stem that allows air to be pumped into a tire and prevents it from flowing out.

_____ **20.** The outside edge of the wheel that helps keep the tire from popping off the wheel.

Multiple Choice

Read each item carefully, and then select the best response.

_____ **1.** The force that acts between the tread and the road surface is called _____.
  A. side force
  B. road force
  C. cornering force
  D. slip angle

_____ **2.** When the front slip angles are larger than the rear slip angles, the vehicle is said to be in a(n) _____ condition, which is referred to as the vehicle "pushing" in the corners.
  A. understeer
  B. oversteer
  C. neutral steer
  D. lateral steer

_____ **3.** The distance across the rim flanges at the bead seat is called the _____.
  A. drop center
  B. flange width
  C. contact area
  D. rim width

_____ **4.** What type of rim is designed to hold the tire bead in place on the rim in the event of inadequate tire pressure?
  A. Split rims
  B. Drop well rims
  C. Safety rims
  D. Steel-disc type

_____ **5.** All of the following are standard types of wheel retaining studs or nuts, _except_:
  A. tapered seat
  B. tapered seat with washer
  C. flat seat without washer
  D. flat seat with washer

_____ **6.** When the plane of the hub mounting surface is shifted from the centerline toward the outside or front side of the wheel it has a _____.
  A. positive offset
  B. negative offset
  C. positive caster
  D. negative camber

_____ **7.** The measure of the innate strength of a material is called _____.
  A. ply rating
  B. tensile strength
  C. flexibility
  D. elasticity

_____ **8.** The first two letters following the letters "DOT" on a tire identify the _____.
  A. tire manufacturer
  B. tire's size
  C. manufacturing plant
  D. both A and C

_____ **9.** The standardized system designed to provide tire buyers with a comparative measure of a tire's tread life, traction, and temperature characteristics, is known as the _____.
- **A.** Uniform Tire Quality Grading System
- **B.** Department of Transportation compliance code
- **C.** International Organization for Standardization Tire Class
- **D.** Manufacturer's load index

_____ **10.** If a tire has a four-digit Department of Transportation code, when was it manufactured?
- **A.** 1970s
- **B.** 1980s
- **C.** 1990s
- **D.** After 2000

_____ **11.** What type of monitoring system utilizes the wheel speed antilock braking system to measure the difference in the rotational speed of the four wheels?
- **A.** Direct TPMS
- **B.** Centrifugal TPMS
- **C.** Indirect TPMS
- **D.** Automatic TPMS

_____ **12.** What type of tires are designed to get the driver to a service center in order to have the regular tire fixed or purchase a new one?
- **A.** Space-saver
- **B.** Self-sealing
- **C.** Run-flat
- **D.** Zero-pressure

_____ **13.** The amount of air pressure in the tire that provides it with load-carrying capacity and affects the overall performance of the vehicle is called the _____.
- **A.** maximum inflation pressure
- **B.** pressure index
- **C.** tire inflation pressure
- **D.** iso pressure

_____ **14.** What type of rotation pattern should you use for a four-tire rotation with nondirectional tires?
- **A.** Forward-cross
- **B.** Rearward-cross
- **C.** X pattern
- **D.** Any of the above

_____ **15.** What type of balancing does not take into consideration that the tire has width?
- **A.** Dynamic
- **B.** Road force
- **C.** Static
- **D.** Horizontal

True/False

If you believe the statement to be more true than false, write the letter "T" in the space provided. If you believe the statement to be more false than true, write the letter "F".

_____ **1.** During cornering, centrifugal force puts more weight on the outside wheels.

_____ **2.** The bead seat is the edge of the rim that creates a seal between the tire bead and the wheel.

_____ **3.** Vehicles manufactured in the mid-70s through the 80s were typically built with positive offset wheels.

_____ **4.** Most lug nuts and studs are left-hand threaded, which means they tighten when turned clockwise.

_____ **5.** Pick-up trucks and large SUVs can have as many as 6, 8, or 10 studs.

_____ **6.** The primary function of the Schrader valve is to assist in keeping debris out of the valve stem.

_____ **7.** Bias-ply tires have much more flexible sidewalls because they use two or more layers of casing plies.

_____ **8.** A properly inflated radial tire runs cooler than a comparable bias-ply tire.

_____ **9.** Asymmetric tread patterns are designed in such a way that the tire can be mounted on the wheel for any direction of rotation.

_____ **10.** Radial tires are marked with the section width in millimeters, but with the rim diameter in inches.

_____ **11.** The lower a tire's aspect ratio, the wider the tire is in relation to its height.

_____ **12.** The DOT tire date manufacturing code is a six-digit code.

_____ **13.** It is illegal to sell a tire intended for use on a public road within the United States without a DOT stamp.

_____ **14.** The use of a centrifugal switch in the indirect TPMS sensor allows the sensor to go to sleep when the vehicle stops.

_____ **15.** Self-sealing tires feature standard tire construction with the addition of a flexible and malleable lining inside the tire in the tread area.

_____ **16.** The proper tire pressure for the vehicle is referred to as the recommended cold inflation pressure.

_____ **17.** Nitrogen-filled tires can be identified by the green valve stem caps that are placed on the valve stem when the tire is inflated.

_____ **18.** Directional tires require keeping the tires on the same side of the vehicle during rotation and generally use a front-to-rear pattern.

_____ **19.** Dynamic imbalance tends to cause the tire to move purely up and down.

_____ **20.** A tire pressure gauge measures pressures in pounds per square inch (psi), kilopascals (kPa), or bars.

Fill in the Blank

Read each item carefully, and then complete the statement by filling in the missing word(s).

1. _____ is when the rear slip angle is larger than the front slip angle; it is referred to as the vehicle being "loose" in the corners.

2. The _____ of a wheel is the outer circular lip of the metal on which the inside edge of the tire is mounted.

3. In _____ _____ wheels, the drop center is closer to the rear of the wheel.

4. Wheels are fastened to the rims by wheel _____ and _____ nuts.

5. The _____ of a wheel is the distance from its hub mounting surface to the centerline of the wheel.

6. After the wheel assembly has been fitted, the wheel retaining nuts that hold the wheel onto the vehicle are _____ to the manufacturer's specifications.

7. The _____ _____ refers to the number and spacing of the lug nuts or wheel studs on the wheel hub on the wheel rim.

8. A(n) _____ _____ is a specially designed opening that allows a tire to be inflated and then automatically closes to prevent air from escaping.

9. A(n) _____-_____ tire is the older form of tire and is still in use on some trailers and off-road vehicles, primarily because of a slightly lower cost and their more durable construction.

10. Directional _____ _____ are designed to provide a range of attributes during particular driving conditions.

11. _____ tread patterns have the same tread pattern on both sides of the tire and are usually nondirectional tires.

12. _____ _____ is a representation of a tire's ability to resist and dissipate heat.

13. Tires with a speed rating designation of _____ may be driven up to 124 mph (200 kph).

14. The _____ _____ of a tire is the ratio of its height to its width. It is usually given as a percentage.

15. As part of US _____ _____ _____ regulations, there must be a tire manufacture date code stamped on the sidewall of every tire.

16. A(n) _____ _____ _____ system monitors the tires for low air pressure and alerts the driver when one or more tires are lower than (or in some cases, higher than) the designated thresholds.

17. The major safety benefit of _____-_____ technology is that it enables a driver to maintain vehicle control if a tire suffers a rapid pressure loss when in motion.

18. The _____ _____ is inspected to ensure that it is above the built-in wear indicators, which indicate that the tire is at the end of its legal life and should be replaced.

19. Filling the tires with _____ will increase the vehicle's fuel efficiency and tire life by reducing the effects of pressure loss due to permeation.

20. _____ balancing is performed by placing specific amounts of weight on each side of the rim to provide the exact counterbalance needed.

Labeling

Label the following diagrams with the correct terms.

1. Parts of a tire:

A. _____

B. _____

C. _____

D. _____

E. _____

F. _____

G. _____

H. _____

I. _____

2. Parts of a radial tire:

A. _____

B. _____

C. _____

D. _____

E. _____

F. _____

G. _____

H. _____

I. _____

J. _____

K. _____

Crossword Puzzle

Use the clues to complete the puzzle.

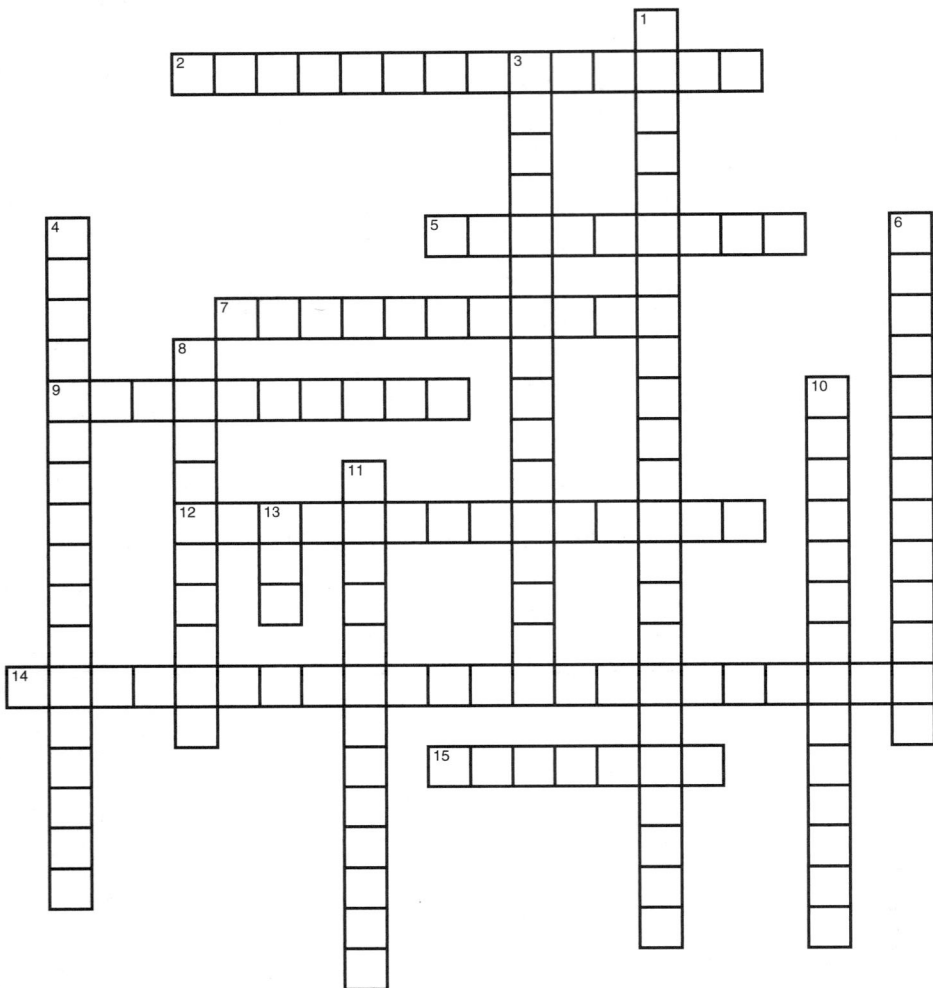

Across

2. The process of matching the tire's lightest point with the rim's heaviest point (generally at the valve stem) for the purpose of reducing the tire's radial imbalance.

5. A condition in which a vehicle's front slip angles are larger than the rear slip angles. This vehicle is said to be "pushing" in the corners.

7. The ratio of sidewall height to section width of a tire.

9. A tire with two or more layers of casing plies and cord loops running radially from bead to bead.

12. The force between the tread and the road surface as a vehicle turns.

14. A tread pattern that differs on each side and, therefore, is usually directional.

15. Nuts that secure the wheel onto the wheel studs.

Down

1. A tread pattern designed to pump water out from under the tire; each tire must be placed in a particular spot on the vehicle.

3. A standardized grading system that indicates the extent to which heat is generated or dissipated by a tire.

4. A switch that is only activated when centrifugal forces are placed on a vehicle.

6. The process of matching the tire's highest point with the rim's lowest point for the purpose of reducing the tire's radial runout.

8. A type of automated tire pressure monitoring system that measures tire pressure and possibly temperature via a sensor installed inside each wheel.

10. A condition in which the plane of the hub mounting surface is positioned toward the outside or front of the wheel centerline.

11. A wheel with negative offset, which gives the outside of the wheel a deep dish appearance.

13. The outer circular lip of the metal on which the inside edge of the tire is mounted.

ASE-Type Questions

Read each item carefully, and then select the best response.

_____ **1.** Tech A says that when turning a corner, both wheels being steered remain parallel to each other as the wheels are steered. Tech B says that on some vehicles, the rear wheels also can be steered. Who is correct?
   A. Tech A
   B. Tech B
   C. Both A and B
   D. Neither A nor B

_____ **2.** Tech A says that most wheels have a drop center or deep well that is used in installing a tire on the wheel. Tech B says that the drop center or deep well is used to prevent the tire from coming off the wheel in the case of low tire pressure. Who is correct?
   A. Tech A
   B. Tech B
   C. Both A and B
   D. Neither A nor B

_____ **3.** Tech A says that all wheels must be torqued to prevent wheels from loosening up and falling off. Tech B says that all wheels must be torqued to prevent overtightening, which can weaken lug studs and warp brake rotors. Who is correct?
   A. Tech A
   B. Tech B
   C. Both A and B
   D. Neither A nor B

_____ **4.** Tech A says that underinflated tires reduce fuel economy. Tech B says that underinflated tires are unsafe and cause accelerated tire wear. Who is correct?
   A. Tech A
   B. Tech B
   C. Both A and B
   D. Neither A nor B

_____ **5.** Tech A says that incorrect toe settings will cause feathered wear across the tire tread. Tech B says that underinflated tires have more wear in the center of the tread. Who is correct?
   A. Tech A
   B. Tech B
   C. Both A and B
   D. Neither A nor B

_____ **6.** Tech A says that tires are marked with a date code indicating the date the tires should be discarded. Tech B says that any tires with a three-digit date code should be discarded. Who is correct?
   A. Tech A
   B. Tech B
   C. Both A and B
   D. Neither A nor B

_____ **7.** Tech A says that a TPMS system can save fuel over time. Tech B says that a TPMS will help prevent blowouts. Who is correct?
   A. Tech A
   B. Tech B
   C. Both A and B
   D. Neither A nor B

_____ 8. Tech A says that the use of nitrogen to fill a tire will prevent the tire from blowing out. Tech B says that when mounting or dismounting a tire on a wheel with a TPMS sensor, you need to position the wheel/tire properly on the tire machine or the TPMS sensor can be easily broken. Who is correct?
  A. Tech A
  B. Tech B
  C. Both A and B
  D. Neither A nor B

_____ 9. Tech A says that using a tire plug to repair a hole in a tire is the best and fastest way to fix a tire. Tech B says that using a tire plug patch is the only approved method of repairing a tire in many states. Who is correct?
  A. Tech A
  B. Tech B
  C. Both A and B
  D. Neither A nor B

_____ 10. Tech A says that directional tires cannot be rotated. Tech B says that all old wheel weights should be removed before balancing a tire. Who is correct?
  A. Tech A
  B. Tech B
  C. Both A and B
  D. Neither A nor B

Servicing the Steering System

Chapter Review

The following activities have been designed to help you refresh your knowledge of this chapter. Your instructor may require you to complete some or all of these activities as a regular part of your training program. You are encouraged to complete any activity that your instructor does not assign as a way to enhance your learning.

Matching

Match the following terms with the correct description or example.

A.	Active control	**K.**	Power section
B.	Actuating	**L.**	Power unit
C.	Beam axle	**M.**	Rack
D.	Chassis	**N.**	Relay lever
E.	Drag link	**O.**	Spline
F.	Gear reduction	**P.**	Steering box
G.	Inertia	**Q.**	Steering linkage
H.	Intermediate shaft	**R.**	Steering system
I.	Knuckle	**S.**	Tie-rod
J.	Pitch	**T.**	Torque sensor

_____ **1.** Steel rods that connect the steering box to the steering arms on the steering knuckle.

_____ **2.** A steel rod that transfers movement from the drag link to an idler arm.

_____ **3.** A device used to measure the load on the steering wheel.

_____ **4.** A chamber in the rack where pressurized fluid acts upon pistons that assist in steering.

_____ **5.** A steel or iron rod that transfers movement of the pitman arm to a relay lever.

_____ **6.** A device that converts the rotary motion of the steering wheel to the linear motion needed to steer the vehicle.

_____ **7.** A suspension system in which one set of wheels is connected laterally by a single beam or shaft.

_____ **8.** A steering component that transfers linear motion from the steering box to the steering arms at the front wheels.

_____ **9.** A system of providing constant feedback from sensors in the vehicle to the control unit.

_____ **10.** A term used to describe all of the components and parts involved in steering a vehicle.

_____ **11.** The act of making something move or work.

_____ **12.** The part that contains the wheel hub or spindle and attaches to the suspension components.

_____ **13.** A gear ratio used to make large turns of the steering wheel into smaller turns of the tire to ease steering for the driver.

_____ **14.** A belt- or gear-driven pump that produces hydraulic pressure for use in the steering box or rack.

_____ **15.** The resistance to a change in motion.

_____ **16.** A ridge or tooth on a driveshaft that meshes with grooves in a mating piece and transfers torque to it, maintaining the angular correspondence between them.

_____ **17.** The frame of a vehicle, to which the suspension pieces attach.

_____ **18.** On a helix, the distance moved in one full revolution of the cylinder.

_____ **19.** A steel rod positioned at an angle from the steering column to the steering gear that functions in transferring movement from one to the other.

_____ **20.** A steel rod driven by the pinion with tie-rods on each end or tie-rods connected to the center of the rack.

Multiple Choice

Read each item carefully, and then select the best response.

_____ **1.** Parallelogram steering uses a _____ gearbox, which changes the direction of steering wheel rotation 90 degrees.
   A. pitman
   B. rack and pinion
   C. worm
   D. linear

_____ **2.** What type of steering system is used on the majority of front-wheel drive vehicles due to the restriction of space under the hood?
   A. Worm and sector
   B. Rack and pinion
   C. Recirculating ball
   D. Direct linkage

_____ **3.** As the driver turns the steering wheel the forces are transferred to the _____, causing the rack to move in either direction.
   A. worm gear
   B. pitman arm
   C. tie-rods
   D. pinion

_____ **4.** The _____ is attached between the tie-rod shaft and the steering arm; it pivots as the rack is extended or retracted when the vehicle is negotiating turns.
   A. inner tie-rod
   B. outer tie-rod
   C. pitman arm
   D. idler arm

_____ **5.** The _____ connects the pitman arm to the idler arm.
   A. center link
   B. tie-rod
   C. steering knuckle
   D. adjustment sleeve

_____ **6.** The _____ connects the tie-rod to the tie-rod end and provides the adjustment point for setting toe-in or toe-out.
   A. adjustment sleeve
   B. idler arm
   C. pitman arm
   D. center link

_____ **7.** The rotary electrical connector located between the steering wheel and the steering column that maintains a constant electrical connection with the wiring while the vehicle's steering wheel is being turned is called the _____.
   A. swivel switch
   B. air bag relay
   C. slot switch
   D. clock spring

_____ **8.** In the rack-and-pinion steering system, the steering rack is supported at the pinion end by being sandwiched between the pinion and a(n) _____.
   A. flexible rubber bellows
   B. spring-loaded rack guide yoke
   C. idler arm
   D. pitman arm

———— **9.** Nylon bushings enable the flexibility needed to accommodate slight radial or lateral movement to assist in ————, or release when something gets stuck.

    A. springback

    B. antibinding

    C. slippage

    D. torsion

———— **10.** The worm-and-nut steering gear is also known as a ———— steering gear.

    A. recirculating ball

    B. box-type

    C. worm and sector

    D. rack and pinion

———— **11.** When the rear wheels can be steered independently of or in conjunction with the front wheels it is known as ————.

    A. independent steering

    B. four-wheel drive

    C. four-wheel steering

    D. worm and roller steering

———— **12.** All of the following are types of power steering, *except*:

    A. hydraulically assisted power steering

    B. compressed air-assisted power steering

    C. electrically powered hydraulic steering

    D. fully electric power steering

———— **13.** The spring-loaded piece of steel connected to the pinion gear or worm at its bottom end and the input shaft at its top end is called a ————.

    A. clock spring

    B. helix

    C. leaf spring

    D. torsion bar

———— **14.** All power steering pumps have a(n) ———— to vary fluid flow and power steering system pressures.

    A. pressure relief valve

    B. spool valve

    C. actuator

    D. flow-control valve

———— **15.** What type of steering system still uses fluid and a pump, but the pump is driven by an electric motor to reduce power drawn from the engine?

    A. Electrically assisted steering system

    B. Electric power steering system

    C. Electrically powered hydraulic steering system

    D. Hybrid steering system

True/False

If you believe the statement to be more true than false, write the letter "T" in the space provided. If you believe the statement to be more false than true, write the letter "F".

———— **1.** The steering box converts the rotary motion of the steering wheel to the linear motion needed to pivot the wheels.

———— **2.** The parallelogram steering system gets its name because the center link and axle, along with the pitman arm and idler arm, always move parallel to each other.

———— **3.** The outer tie-rod is attached to the end of the rack and allows for suspension movement and slight changes in steering angles.

———— **4.** Toe-out is a condition where the fronts of the wheels, as seen from above, are closer together than the rears of the wheels.

———— **5.** The toe-setting is the symmetric angle that each wheel makes with the longitudinal axis of the vehicle.

_____ **6.** To reduce serious injury, all steering columns are now fitted with collapsible sections that help protect the driver.

_____ **7.** When a technician needs to carry out a servicing procedure on the steering column, it is good practice to disarm the triggering system for the driver's side airbag.

_____ **8.** Both the pinion and the rack teeth are worm gears.

_____ **9.** A disadvantage of the rack-and-pinion steering system is that it typically is not manually adjustable.

_____ **10.** Worm gear steering boxes made the process of turning the front wheels an easier task for drivers of the early automobile.

_____ **11.** A ball-return guide is a special passage or metal tube through which the balls move in recirculating ball steering boxes.

_____ **12.** Power steering fluid gets contaminated with rubber and metallic particles from internal wear in the system.

_____ **13.** Direct drive steering is a completely electrically powered power-assist system that eliminates all hydraulic components and fluid.

_____ **14.** In a pinion-assist type EPS steering system, the power assist unit, controller, and torque sensor are attached to the steering column.

_____ **15.** In an EPS system, sensor inputs are compared to determine how much power assistance is required according to the forces capability map data stored in the ECU's memory.

Fill in the Blank

Read each item carefully, and then complete the statement by filling in the missing word(s).

1. The _____ _____ transmits the driver's steering effort from the steering wheel down to the steering box.

2. The steering linkage transfers the linear steering effort to the wheels by connecting the steering box to the _____ _____ on each of the steering knuckles, which pivot on the ball joints, allowing the wheels to steer the vehicle.

3. The steering knuckle pivots on one or two _____ _____, depending on the type of suspension.

4. The relationships between the steering system, the wheel positions, and the suspension system form what is called the _____ _____.

5. The rack slides in the housing and is moved by the action of the _____ pinion.

6. The _____ _____ protects the inner joints from dirt and contaminants and retains the grease lubricant inside the rack-and-pinion housing.

7. The _____ _____ transfers movement from the steering box to the center link.

8. The _____ _____ is attached to the chassis (the frame of the vehicle) and is positioned parallel to the pitman arm.

9. _____ _____ is the undesired condition produced when, upon hitting a bump, the vehicle darts to one side as the steering linkage is pushed or pulled as a result of the travel of the suspension.

10. To compensate for variations in driving positions, many manufacturers have included a steering column _____ and/or _____ mechanism to their vehicles.

11. Mechanical advantage is gained in the steering system by the _____ _____ between the turns of the steering wheel and the angle turned of the wheel.

12. The rack is typically supported at both ends of the _____ _____ or tube by a nylon bushing.

13. The pinion is supported by two bearings in the rack housing that must be _____ to ensure the pinion is in the correct position, relative to the rack, and to eliminate free play.

14. The steering arm, the stub axle knuckle, and the stub-axle carrier can be forged as one piece, and can be referred to as a(n) _____ _____.

15. The _____ _____ can read both torque and rotation from the steering wheel and convert them into voltage signals for the ECU to monitor.

Labeling

Label the following diagrams with the correct terms.

1. The components of a basic steering system:

A. _____

B. _____

C. _____

D. _____

E. _____

F. _____

G. _____

H. _____

I. _____

J. _____

K. _____

L. _____

M. _____

2. Rack-and-pinion steering system:

A. _____

B. _____

C. _____

D. _____

E. _____

F. _____

G. _____

H. _____

I. _____

J. _____

3. Recirculating ball steering system:

A. _____

B. _____

C. _____

D. _____

E. _____

F. _____

G. _____

4. Power-assisted rack-and-pinion system:

A. _____

B. _____

C. _____

D. _____

E. _____

F. _____

G. _____

H. _____

I. _____

J. _____

K. _____

L. _____

M. _____

N. _____

O. _____

P. _____

Q. _____

5. Power-assisted recirculating gearbox:

A. _____

B. _____

C. _____

D. _____

E. _____

F. _____

Crossword Puzzle

Use the clues to complete the puzzle.

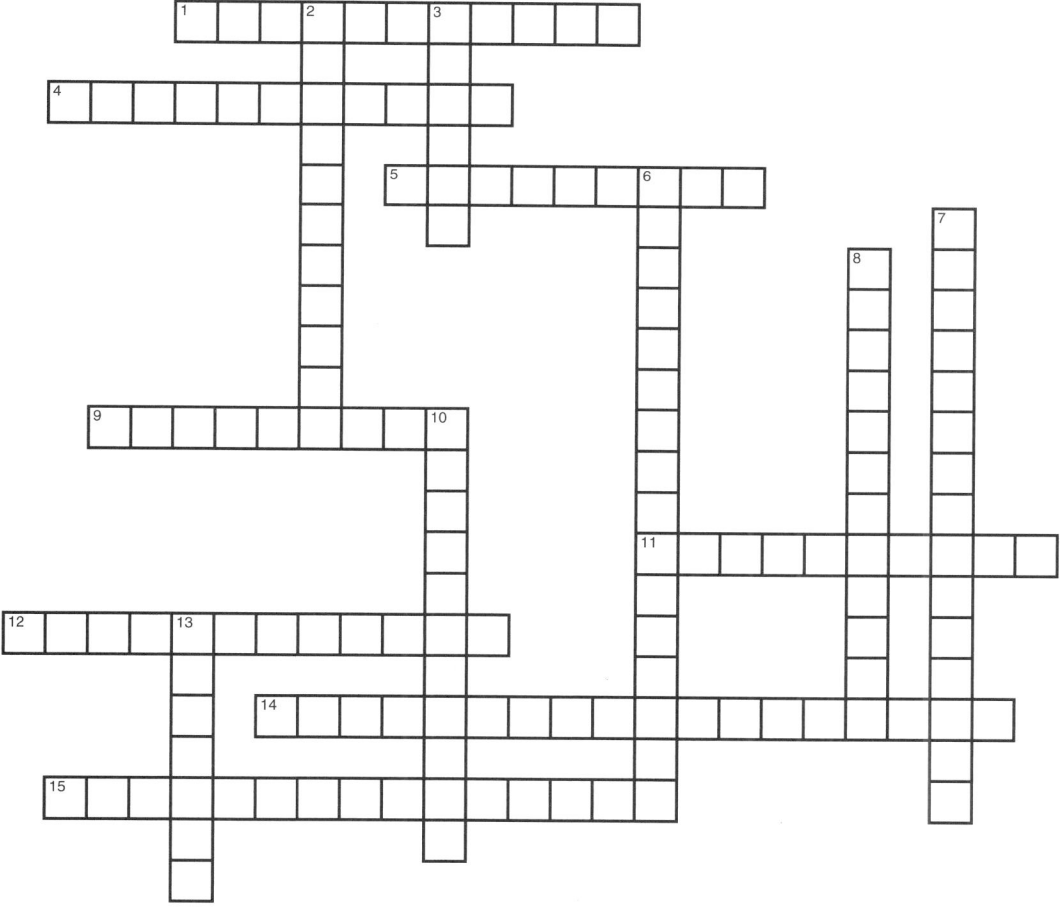

Across

1. A special rotary electrical connector located between the steering wheel and the steering column that maintains a constant electrical connection with the wiring system while the vehicle's steering wheel is being turned.

4. The release of something when it gets stuck.

5. The protrusion of the worm gear that serves as the point of attachment to the steering column.

9. The undesired condition produced when hitting a bump where the vehicle darts to one side as the steering linkage is pushed or pulled as a result of the travel of suspension.

11. Setting of the toe-in or toe-out of the tires to the centerline of the vehicle.

12. A pinion when it is mated with the rack.

14. A brace or nylon part that pushes against the rack to adjust the mesh of the rack teeth to the pinion teeth.

15. A special passage or metal tube through which the balls move in recirculating ball steering boxes.

Down

2. Any device that controls another object such as a computer.

3. A gear located at the end of the steering shaft connected to the rack. It moves from side to side as the pinion rotates, controlling the direction of the wheels.

6. A component that connects the tie-rods together and to the center link on some applications, providing the adjustment point for toe-in or toe-out, depending on the manufacture's specifications.

7. The electric motor in electric power assist steering systems.

8. A term used to describe two or more parts, such as gears, that are in constant contact with each other.

10. The outer shell of the rack-and-pinion steering system that is mounted to the chassis.

13. Unwanted lateral movements of the worm shaft.

ASE-Type Questions

Read each item carefully, and then select the best response.

1. Tech A says that the steering column uses one or more flexible joints to connect to the steering gearbox. Tech B says that the pitman arm is bolted to the frame and relays the steering linkage movement to the opposite wheel from the steering gearbox. Who is correct?
   A. Tech A
   B. Tech B
   C. Both A and B
   D. Neither A nor B

2. Tech A says that the clock spring assists the turning of the steering wheel on vehicles with EPS. Tech B says that the clock spring is used to transmit an electrical signal to the driver's side airbag. Who is correct?
   A. Tech A
   B. Tech B
   C. Both A and B
   D. Neither A nor B

3. Tech A says that power steering fluid is universal and can be used in virtually any vehicle's power steering system. Tech B says that in a power steering system, the force needed to turn the wheels is created by the hydraulic pump or electric motor. Who is correct?
   A. Tech A
   B. Tech B
   C. Both A and B
   D. Neither A nor B

4. Tech A says that rack-and-pinion steering systems generally do not use power steering due to their lighter duty construction. Tech B says that rack-and-pinion steering systems use a worm gear arrangement to move the rack. Who is correct?
   A. Tech A
   B. Tech B
   C. Both A and B
   D. Neither A nor B

_____ **5.** Tech A says that to properly check a tie-rod end, the technician should twist the tie-rod end, and any rotational movement means the joint is bad. Tech B says that to properly check tie-rod ends, the vehicle's weight should be on the wheels, and as an assistant wiggles the steering wheel, you can check for movement in the tie-rod ends. Who is correct?

  A. Tech A

  B. Tech B

  C. Both A and B

  D. Neither A nor B

_____ **6.** Tech A says that worn rack-and-pinion mount bushings can cause excessive play in the steering system. Tech B says that a worn idler arm can cause excessive play in the steering system. Who is correct?

  A. Tech A

  B. Tech B

  C. Both A and B

  D. Neither A nor B

_____ **7.** Tech A says that when the vehicle is being driven straight ahead, a belt-driven power steering pump will pump fluid continuously, placing a minimal load on the engine. Tech B says that a belt-driven power steering pump is activated by an electromagnetic clutch, so it only pumps fluid when the wheels are being steered. Who is correct?

  A. Tech A

  B. Tech B

  C. Both A and B

  D. Neither A nor B

_____ **8.** Tech A says that the pressure relief valve maintains a preset minimum pressure in the system. Tech B says that the pressure relief valve prevents excessive pressure. Who is correct?

  A. Tech A

  B. Tech B

  C. Both A and B

  D. Neither A nor B

_____ **9.** Tech A says that worn tie-rod ends can cause a steering wandering complaint. Tech B says that a hard steering complaint could be caused by a worn power steering pump. Who is correct?

  A. Tech A

  B. Tech B

  C. Both A and B

  D. Neither A nor B

_____ **10.** Tech A says that a pickle fork is used to hold a tie-rod while it is being tightened. Tech B says that the pitman arm is usually threaded so that the front wheel toe can be adjusted. Who is correct?

  A. Tech A

  B. Tech B

  C. Both A and B

  D. Neither A nor B

Servicing the Suspension System

Chapter Review

The following activities have been designed to help you refresh your knowledge of this chapter. Your instructor may require you to complete some or all of these activities as a regular part of your training program. You are encouraged to complete any activity that your instructor does not assign as a way to enhance your learning.

Matching

Match the following terms with the correct description or example.

A. Axle
B. Caster
C. Deflecting force
D. Elasticity
E. Independent suspension
F. Negative camber
G. Overshoot
H. Positive caster
I. Rebound clip
J. Scrub radius

K. Shroud
L. Sintering
M. Static toe
N. Suspension strut
O. Suspension system
P. Thrust line
Q. Torsion bar
R. Trailing arm suspension
S. Uniform pitch
T. Yaw

_____ 1. A system within a vehicle designed to isolate the vehicle body from road bumps and vibrations.

_____ 2. A steel or plastic cover placed over the shock rod.

_____ 3. A spring whose pitch (the distance from the center of one coil to the center of the adjacent coil) is the same distance throughout.

_____ 4. The angle formed through the wheel pivot points when viewed from the side in comparison to a vertical line through the wheel.

_____ 5. Movement of a vehicle around its z-axis (vertical axis) felt when the vehicle deviates from its straight path, as when skidding sideways and the rear comes around.

_____ 6. Backward tilt of the wheel pivot points from the vertical line.

_____ 7. The shaft of the suspension system to which the tires and wheels are attached; they are used to drive or support the wheels.

_____ 8. A metal strap that is warped around the leaf spring to prevent excessive flexing of the main leaf during rebound.

_____ 9. The ability to deform and reform into the same shape.

_____ 10. The process of using pressure and heat to bond metal particles.

_____ 11. The amount a spring extends (springs back) past its original length following compression.

_____ 12. A setting designed to compensate for slight wear in steering components that may cause the wheels to turn outward or inward while the vehicle is in motion.

_____ 13. Tilt of the top of the tire toward the centerline of the vehicle.

_____ 14. The imaginary line drawn through the center of the rear axle.

_____ 15. A force that moves an object in a different direction or into a different shape.

_____ 16. A bar made of a steel alloy that is fixed rigidly to the chassis at one end and the suspension control arm at the other to support the weight of a vehicle.

_____ 17. The distance between two imaginary points on the road surface—the point of center contact between the road surface and the tire, and the intersecting point where the steering axis centerline and the tire centerline contact the road surface.

_____ 18. A type of suspension system that uses upper and lower control arms.

_____ 19. A shock absorber designed to reduce spring oscillations.

_____ 20. A system for allowing the up and down movement of one tire without affecting the other tire on that axle.

Multiple Choice

Read each item carefully, and then select the best response.

_____ 1. A force that moves an object in a different direction or into a different shape is known as a(n) _____.
   A. applied force
   B. reaction force
   C. deflecting force
   D. suspension force

_____ 2. Parts of a vehicle that are not supported by the suspension system, including the wheels, tires, brakes, axles, and steering and suspension parts not supported by springs are considered _____.
   A. unsprung mass
   B. free weight
   C. unsprung weight
   D. either A or C

_____ 3. The force transferred from the tire contact patch through the axle housing is called _____.
   A. cornering force
   B. driving thrust
   C. reaction force
   D. roll

_____ 4. Movement around the vehicle's y-axis, commonly felt during hard braking or fast acceleration, when the front of the vehicle noses down or rises up slightly, is called _____.
   A. yaw
   B. pitch
   C. roll
   D. oscillation

_____ 5. When a spring deflects easily under a light load, but its resistance increases as the load increases, it is said to have a(n) _____.
   A. progressive rate of deflection
   B. constant rate of deflection
   C. uniform pitch
   D. progressive pitch

_____ 6. The rear of the multileaf is connected to the frame by a(n) _____, which provides a link between the spring eye and a bracket on the frame.
   A. rigid spring hanger
   B. intermediate hook
   C. rebound clip
   D. swinging shackle

_____ 7. The twisting force that is applied by anchoring one end of an object and then applying a twisting force to the other end is called _____.
   A. torsional load
   B. compression
   C. pitch
   D. deflection

_____ 8. A(n) _____ consists of a flexible rubber bladder, which seals the outside of the upper and lower halves of the shock absorber.
   A. strut-type shock absorber
   B. air spring
   C. gas-pressurized shock absorber
   D. suspension strut

_____ 9. The newest style of electronic adjustable shock uses a special type of fluid called _____ fluid that has the unique characteristic of changing viscosity when exposed to a magnetic field.
   A. anisotropic fluid
   B. magnetic hydraulic
   C. magneto-rheological
   D. frequency

_____ 10. In control arm applications, particularly at the rear of a vehicle, a rubber bushing may be molded with a voided section; this bushing is called a _____.
   A. rubber-bonded bushing
   B. spring shackle bushing
   C. gap-tooth bushing
   D. compliance bushing

_____ 11. What is another name for a rigid-axle coil-spring suspension that uses two bars similar to a panhard rod and a pivot point on the axle to keep the axle from moving in turns?
   A. Watt's linkage
   B. Rigid nondrive axle suspension
   C. Trailing arm suspension
   D. Rear-wheel independent suspension

_____ 12. Which of the following is an example of a computerized suspension system?
   A. Stepper motor actuated
   B. Solenoid valve actuated
   C. Electromagnetic rheological
   D. All of the above

_____ 13. The side-to-side vertical tilt of the wheel, viewed from the front of the vehicle and measured in degrees is called _____.
   A. caster
   B. camber
   C. alignment
   D. steering axis

_____ 14. When a vehicle has _____, a line drawn through the steering axis centerline meets the road surface behind the vertical centerline of the wheel.
   A. positive camber
   B. negative camber
   C. positive caster
   D. negative caster

_____ 15. The distance between two imaginary points on the road surface is called _____.
   A. scrub radius
   B. steering offset
   C. scrub geometry
   D. all of the above

_____ 16. If the camber line is outside of the steering axis inclination line, then it has _____.
   A. negative offset
   B. zero scrub radius
   C. positive scrub radius
   D. toe-out on turns

_____ **17.** The condition in which the fronts of the wheels are closer together than the rears of the wheels is called _____.
  A. toe-in
  B. toe-out
  C. static toe
  D. none of the above

_____ **18.** The _____ refers to the relationship between the centerline of the vehicle and the angle of the rear tires.
  A. thrust line
  B. thrust angle
  C. turning radius
  D. Ackermann angle

_____ **19.** The amount of distance between the ground and a specified part of the vehicle such as the fender well, upper control arm, or rocker panel is called _____.
  A. clearance
  B. setback
  C. ride height
  D. camber

_____ **20.** What type of alignment is outdated and almost never performed on modern vehicles?
  A. Front-end two-wheel alignment
  B. Thrust-angle alignment
  C. Four-wheel alignment
  D. None of the above

True/False

If you believe the statement to be more true than false, write the letter "T" in the space provided. If you believe the statement to be more false than true, write the letter "F".

_____ **1.** Preventing or reducing oscillations is called dampening.

_____ **2.** Pitch is vehicular movement along its x-axis. It is the rolling motion you feel when making a sharp corner and is generally what causes rollovers.

_____ **3.** Shock absorbers dampen spring oscillations by forcing water through small holes in a piston.

_____ **4.** The pitch of a spring is the distance from the center of one coil to the center of the adjacent coil; if they are evenly spaced it is called uniform pitch.

_____ **5.** The longest leaf is called the main leaf and is rolled at both ends to form spring eyes; they are used to mount the spring to the frame of the vehicle.

_____ **6.** Leaf springs can be used across the chassis frame in a trailing arm suspension, or as part of the connecting link between two axle assemblies on a semi-rigid axle beam.

_____ **7.** In full-floating axles, the axle bearing is placed on the outside of the axle housing.

_____ **8.** The most widely used hydraulic shock absorber is the direct-acting telescopic type.

_____ **9.** Electronic adjustable-rate shock absorbers are also called self-leveling shock absorbers.

_____ **10.** Typically straight pieces of steel used to either transfer motion or prevent motion within a vehicle's suspension system are called control arms.

_____ **11.** Rubber-bonded bushings are normally used for the front eye of the spring at the fixed shackle point, and also in control arm applications.

_____ **12.** A solid axle is a nonindependent suspension because the wheels on both sides of the axle are connected together.

_____ **13.** MacPherson strut suspension can be used on either the front or the rear of the vehicle.

_____ **14.** The primary reason that the short-/long-suspension system was designed was to ensure correct wheel alignment as the vehicle corners.

_____ **15.** Most twin I-beam systems use a coil spring to support the weight of the vehicle.

_____ **16.** Rear-wheel suspension systems can be of the independent or nonindependent design.

_____ **17.** A wheel that leans away from the center of the vehicle at the top is said to have negative camber.

_____ **18.** If the camber line is outside of the SAI line, then it has positive offset or positive scrub radius.

_____ **19.** The Ackermann angle is the angle the steering arms make with the steering axis, projected toward the center of the rear axle.

_____ **20.** Setback is an alignment angle that is not adjustable but that allows the technician to diagnose the vehicle.

Fill in the Blank

Read each item carefully, and then complete the statement by filling in the missing word(s).

1. When a tire hits an obstruction, there is a _____ _____, meaning the tire will move in response to the force applied by the obstruction.

2. _____ refers to the fluctuating of an object between two states, basically meaning the spring compresses and rebounds over and over again.

3. _____ force refers to the lateral movement of the axle housing during turning.

4. _____ _____ are swivel connections mounted in the outer ends of the control arm.

5. A _____ _____ is made from a single length of special wire, which is heated and wound on a former to produce the required shape.

6. A bar similar to the torsion bar is the _____ _____, or antiroll bar.

7. If the suspension reaches its limit of travel, rubber _____ prevent direct metal-to-metal contact, thereby reducing jarring of the suspension components.

8. A manual adjustable-rate shock absorber has a manual, external _____ _____ adjustment.

9. The A-arm style control arm is sometimes referred to as a _____ control arm, a relatively flat triangular part that mounts to the frame or subframe at each leg of the A.

10. _____ _____ bushings can be molded to form two halves, to fit into each side of the spring eye on the swinging shackle, which is located on the vehicle frame.

11. A(n) _____ _____ allows the wheel on each side of the axle to move up and down independently of each other.

12. In strut suspension systems, the _____ _____ is contained inside the strut.

13. Also referred to as a track bar, a(n) _____ _____ sits parallel with the axle. One end connects to the frame of the vehicle, and the other end connects to the axle.

14. On rear-wheel drive vehicles with independent rear suspension, drive is transmitted to each wheel by external _____ _____.

15. _____ _____ suspension is an electronically controlled air suspension system at all four wheels with a continuously adaptive dampening system.

16. Wheels are positioned on the suspension at certain angles to provide for easy driving of the vehicle. These angles, taken together, determine the vehicle's _____ _____.

17. The angle formed between the steering axis inclination and the camber line is called the _____ _____.

18. _____-_____ on turns is the relative toe setting of the front wheels as the vehicle turns.

19. The _____ _____ is a geometric alignment of linkages in a vehicle's steering such that the wheels on the inside of a turn are able to move in a different circle radius than the wheels on the outside.

20. _____ _____ is a measure of how small a circle the vehicle can turn in when the steering wheel is turned to the limit.

Labeling

Label the following diagrams with the correct terms.

1. The suspension system:

A. _____

B. _____

C. _____

D. _____

E. _____

F. _____

G. _____

H. _____

I. _____

J. _____

2. Yaw, pitch, and roll:

A. _____

B. _____

C. _____

3. Gas-pressurized shock absorber:

A. _____

B. _____

C. _____

D. _____

E. _____

F. _____

G. _____

H. _____

I. _____

J. _____

4. Electromagnetic shock absorber:

A. _____

B. _____

C. _____

D. _____

E. _____

F. _____

G. _____

H. _____

I. _____

5. Automatic load adjustable suspension system:

A. _____

B. _____

C. _____

D. _____

E. _____

F. _____

6. Strut suspension:

A. _____

B. _____

C. _____

D. _____

E. _____

F. _____

G. _____

H. _____

I. _____

J. _____

K. _____

7. Adaptive shock absorber:

A. _____

B. _____

C. _____

D. _____

E. _____

8. Camber:

A. _____

B. _____

C. _____

D. _____

9. Caster:

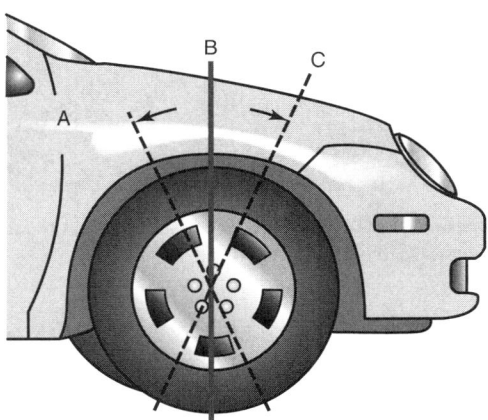

A. _____

B. _____

C. _____

10. Steering axis and scrub radius:

A. _____

B. _____

C. _____

D. _____

Skill Drills

Test your knowledge of skill drills by filling in the correct words in the photo captions.

1. Ride Height:

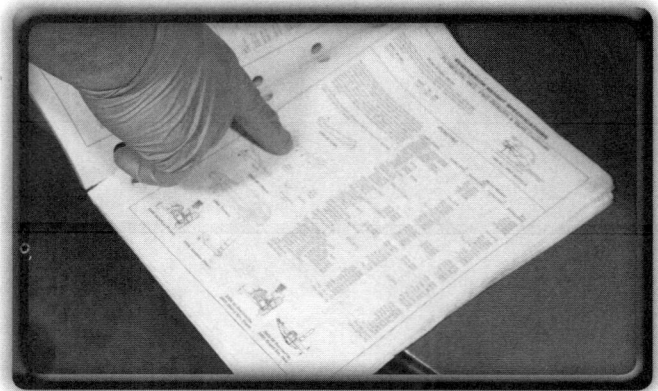

Step 1: Refer to the manufacturer's service information for correct _____ points and _____.

Step 2: Check for properly _____, _____, and _____ tires. Correct any issues found.

Step 3: Check the vehicle for any _____ _____ in the trunk or luggage area. Remove them temporarily while measuring ride height.

Step 4: Measure from points specified, such as from _____ to _____ on all four corners of the vehicle, and _____ measurements to specifications.

Step 5: Inspect for _____ components or a weak or broken spring if any measurements are not correct. If working with a _____ bar suspension, adjustment of ride height may need to be performed to correct the condition.

Step 6: Refer to the manufacturer's service information for proper adjustment of ride height. Remember that anytime ride height is changed, a(n) _____ _____ will need to be performed also. **(There is no image associated with this step.)**

2. Replacing a Shock Absorber:

Step 1: Raise the vehicle on a lift, or raise it with a jack and support it with _____ _____ under the frame. Remove the upper bolts holding the shocks in place with a _____ and box-end wrench.

Step 2: Remove the _____ _____ holding the shock in place.

Step 3: Pull the _____ out by hand. _____ on the other side.

3. Lubricating Suspension and Steering Systems:

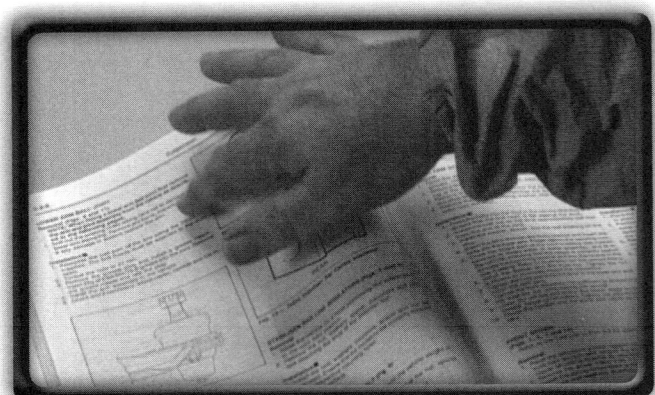

Step 1: Determine the location of lubricating points. Check the shop _____ _____ to determine where the grease points are and the type of grease required. Also look to see if any _____ grease fittings are installed. If so, grease them.

Step 2: Clean each of the lubrication fittings and the grease gun _____ by wiping them with a clean rag. You may need to remove a component's plugs and temporarily install a _____ _____. After the component has been lubricated, reinstall the original plug.

Step 3: Push the _____ _____ nozzle fully over the fitting. It should snap into place. Add enough grease to see the seal or rubber boot rise slightly. Do not _____ a lubricated joint with grease.

Step 4: If the fitting is clean and will not take grease, remove the grease _____ and check for blockage. If found, the fitting must be replaced with a new fitting of the same _____ and _____, and the joint relubricated.

Step 5: Remove the _____ from the _____, and wipe away any excess _____ from it. Repeat the procedure until all the appropriate joints have been lubricated.

Step 6: After you have completed lubricating all the appropriate joints and cleaned off any excess grease, attach a _____ _____ _____ to the windshield, or reset the maintenance reminder system. Lower the vehicle, and remove it from the lifting device.

4. Inspecting Leaf Springs:

Step 1: Raise the vehicle on a lift, and support the _____ _____ with tall screw jack stands. Check to see if any of the _____ are cracked or broken and that the noise _____ inserts are positioned correctly between the leaves.

Step 2: Test the security of the _____ _____ _____, and make sure the U-bolts are tight.

Step 3: Check the condition of the bushings or mountings and the _____ _____ by placing a _____ between the frame and the eye of the spring and levering against the spring.

5. Preparing a Vehicle for a Wheel Alignment:

Step 1: Remove any _____ items from the trunk and _____ compartments.

Step 2: Check the size and _____ of all four tires. Adjust the _____ _____ to specifications.

Step 3: Measure the vehicle's _____ _____.

Step 4: Check the _____ of the steering wheel. Correct any _____ _____ before undertaking the wheel alignment.

Step 5: _____ each corner of the vehicle to check the correct functioning of the _____ _____.

Step 6: With the vehicle raised, inspect all suspension and steering components, including the _____ _____. Repair or replace all damaged or worn _____ _____.

Step 7: Position the vehicle on the wheel alignment _____, making sure the front tires are positioned correctly on the _____.

Step 8: Position the _____ wheels on the _____ _____ or rear turntables.

Step 9: Attach the wheel _____ of the wheel _____ machine.

Crossword Puzzle

Use the clues to complete the puzzle.

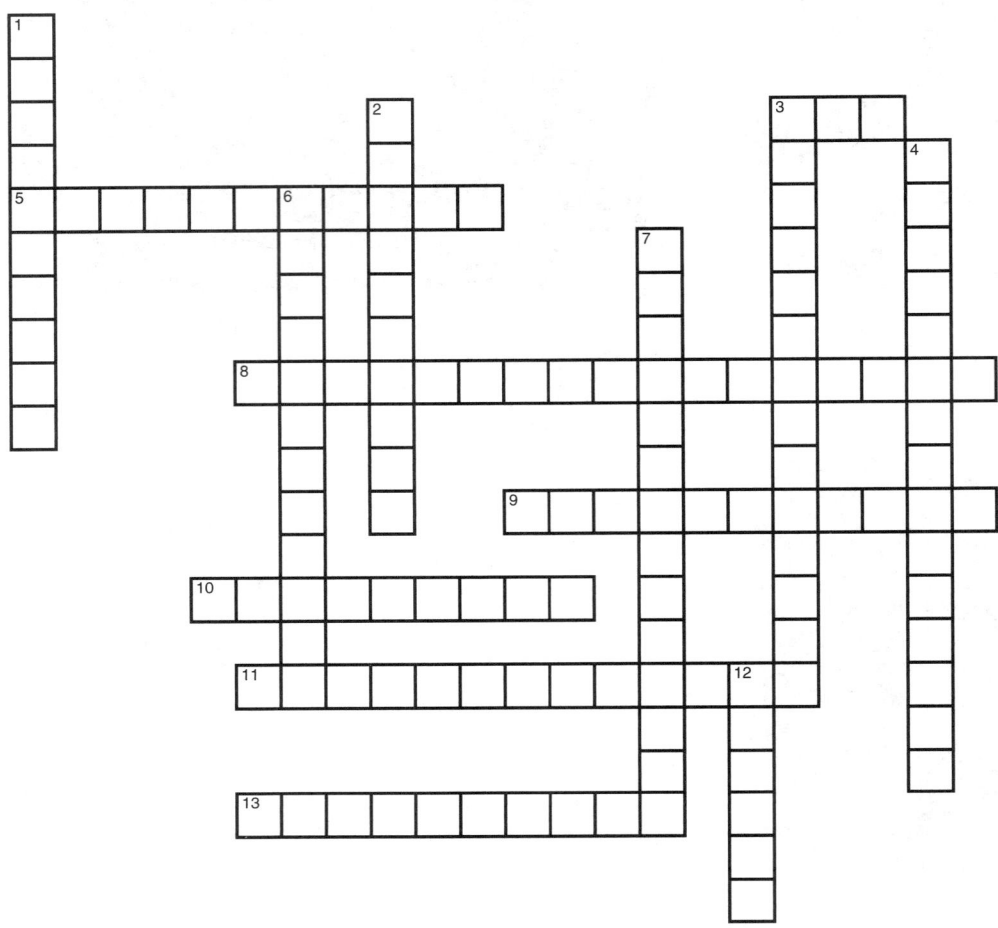

Across

3. The primary load-bearing element of a vehicle's suspension system, commonly known as control arm. Can also mean the steering arm, which applies the driver's turning effort to the steering knuckle.

5. The distance between two imaginary points on the road surface—the point of center contact between the road surface and tire, and the intersecting point where the steering axis centerline and the tire centerline contact the road surface.

8. The rigid part typically welded to the body or frame of the vehicle to which the front of the leaf spring is attached.

9. The fluctuation of an object between two states. With regard to suspension springs, it refers to the uncontrolled compression and decompression of the spring following overshoot.

10. A part that provides the springing action or auxiliary spring. It is typically used in air suspension systems or heavy truck applications.

11. A force that acts in the opposite direction to another force.

13. The imaginary line drawn down the exact center of the vehicle from front to back.

Down

1. A spring made of one or more flat, tempered steel springs bracketed together that is used in the suspension system to support the weight of the vehicle.

2. The amount of ground clearance a vehicle has, measured from a point on the body or frame depending on the manufacturer. Also known as ride height.

3. The angle the steering arms make with the steering axis, projected toward the center of the rear axle.

4. A strut used on an independent suspension where the spring and shock are joined together. Used on most front-wheel drive vehicles.

6. Pressure placed on something.

7. The force applied to an axle that shifts the axle in relation to the position of the body when turning.

12. The side-to-side vertical tilt of the wheel. It is viewed from the front of the vehicle and measured in degrees.

ASE-Type Questions
Read each item carefully, and then select the best response.

_____ 1. Tech A says that the wheels and tires are examples of unsprung weight. Tech B says that the exhaust system is an example of sprung weight. Who is correct?
   A. Tech A
   B. Tech B
   C. Both A and B
   D. Neither A nor B

_____ 2. Tech A says that thrust angle refers to the direction the front wheels are pointing. Tech B says that scrub radius refers to the vertical centerline of the tire in relation to an imaginary line through the steering knuckle pivots. Who is correct?
   A. Tech A
   B. Tech B
   C. Both A and B
   D. Neither A nor B

_____ 3. Tech A says that yaw is when a vehicle deviates from its straight path. Tech B says that roll is felt during hard braking. Who is correct?
   A. Tech A
   B. Tech B
   C. Both A and B
   D. Neither A nor B

_____ 4. Tech A says that a vehicle will tend to pull toward the side with the most positive camber. Tech B says that a progressive-rate spring offers a soft ride but can also carry a heavier load. Who is correct?
   A. Tech A
   B. Tech B
   C. Both A and B
   D. Neither A nor B

_____ 5. Tech A says that a shock dampens movement of the suspension in both upward and downward movement. Tech B says that a shock only dampens upward movement. Who is correct?
   A. Tech A
   B. Tech B
   C. Both A and B
   D. Neither A nor B

_____ 6. Tech A says that ball joints must be unloaded when checking them for wear. Tech B says that ball joints must be loaded when checking them for wear. Who is correct?
   A. Tech A
   B. Tech B
   C. Both A and B
   D. Neither A nor B

_____ 7. Tech A says that positive toe is when the fronts of the tires are farther apart than the rears of the tires. Tech B says that a ball joint in a MacPherson strut suspension is a follower joint (not loaded). Who is correct?
   A. Tech A
   B. Tech B
   C. Both A and B
   D. Neither A nor B

_____ **8.** Tech A says that a dead axle is designed to carry the weight of the vehicle with no drive capability. Tech B says that checking toe-out on turns is a prealignment check. Who is correct?

   A. Tech A
   B. Tech B
   C. Both A and B
   D. Neither A nor B

_____ **9.** Tech A says that loose ball joints can cause the vehicle to wander. Tech B says that loose control arm bushings can affect alignment angles. Who is correct?

   A. Tech A
   B. Tech B
   C. Both A and B
   D. Neither A nor B

_____ **10.** Tech A says that the front wheels should be aligned before the rear wheels. Tech B says that front wheel toe should be adjusted after front wheel camber and caster. Who is correct?

   A. Tech A
   B. Tech B
   C. Both A and B
   D. Neither A nor B

Principles of Braking

Chapter Review

The following activities have been designed to help you refresh your knowledge of this chapter. Your instructor may require you to complete some or all of these activities as a regular part of your training program. You are encouraged to complete any activity that your instructor does not assign as a way to enhance your learning.

Matching

Match the following terms with the correct description or example.

A. Asbestos
B. Band brake
C. Brake assist
D. Brake fade
E. Brakes
F. Conservation of energy
G. Disc brakes
H. Drum brakes
I. Friction
J. Fulcrum

K. Heat fade
L. Kinetic energy
M. Master cylinder
N. Mechanical advantage
O. Mechanical disadvantage
P. Parking brake
Q. Rotational force
R. Scrub brakes
S. Top hat parking brake
T. Weight transfer

_____ **1.** A type of brake system that forces stationary brake pads against the outside of a rotating brake rotor.

_____ **2.** A brake system that uses leverage to force a friction block against one or more wheels.

_____ **3.** When the load distance on a lever is greater than the effort distance, which means the effort required to move the load is greater than the load itself.

_____ **4.** A physical law that states that energy cannot be created or destroyed.

_____ **5.** A drum brake that is located inside a disc brake rotor in order to act as a parking brake.

_____ **6.** A type of brake system that forces brake shoes against the inside of a brake drum.

_____ **7.** An enhanced safety system built in to some ABS systems that anticipates a panic stop and applies maximum braking force to slow the vehicle as quickly as possible.

_____ **8.** The force created by the rotating wheel when the brakes are applied; it causes the brake components to twist the brake support and ultimately the vehicle in the direction of wheel rotation.

_____ **9.** The reduction in stopping power caused by a change in the brake system such as overheating, water, or overheated brake fluid.

_____ **10.** The ratio of load and effort for any simple machine such as a lever.

_____ **11.** A braking system that uses a metal band lined with friction material to clamp around the outside of a wheel or drum.

_____ **12.** The point around which a lever rotates and that supports the lever and the load.

_____ **13.** A mineral with needle-like fibers that can become embedded in lung tissue and cause cancer.

_____ **14.** Weight moving from one set of wheels to the other set of wheels during braking, acceleration, or cornering.

_____ **15.** Brake fade caused by the buildup of heat in braking surfaces, which get so hot they cannot create any additional heat, leading to a loss of friction.

_____ **16.** The resistance created by surfaces in contact.

_____ **17.** Converts the brake pedal force into hydraulic pressure, which is then transmitted via brake lines and hoses to one or more pistons at each wheel brake unit.

_____ **18.** A brake system used for holding the vehicle when it is stationary.

_____ **19.** A system made up of hydraulic and mechanical components designed to slow or stop a vehicle.

_____ **20.** The energy of an object in motion; it increases by the square of the speed.

Multiple Choice

Read each item carefully, and then select the best response.

_____ **1.** The old _____ braking systems caused the vehicle to veer dangerously to one side when braking and required frequent adjustment.
   A. scrub
   B. band
   C. disc
   D. drum

_____ **2.** What type of system uses a computer to monitor each wheel's speed and either hold, decrease, or apply hydraulic pressure to each wheel to prevent wheel lock-up and maintain the maximum amount of braking power just short of brake lock-up?
   A. Brake assist
   B. Antilock brake system
   C. Brake-by-wire
   D. Brake pedal emulator

_____ **3.** In a brake-by-wire system a _____ tells the computer how firmly the driver intends to brake which then sends control signals to the appropriate brake actuators.
   A. brake pedal emulator
   B. service brake
   C. hydraulic actuator
   D. kinetic sensor

_____ **4.** Which of the following factors influence a vehicle's ability to brake effectively?
   A. Weight of the vehicle
   B. Height of the vehicle
   C. Tire composition
   D. All of the above

_____ **5.** What law states that an object will stay at rest or at uniform speed unless it is acted on by an outside force?
   A. Euler's first law of motion
   B. The law of conservation of energy
   C. Newton's first law of motion
   D. Kirchhoff's law

_____ **6.** The amount of friction between two moving surfaces in contact with each other is expressed as a ratio and is called the _____.
   A. law of conservation of energy
   B. coefficient of friction
   C. rate of heat transfer
   D. second law of motion

_____ **7.** When brakes are operated on a moving vehicle, a _____ force is generated.
   A. cornering
   B. directional
   C. deflecting
   D. rotational

_____ **8.** Brake pedals are usually a _____. They pivot at the top end (fulcrum). The foot pressure (effort) is applied to the bottom end. And the master cylinder (load) is applied between the two.
   A. lever of the first order
   B. lever of the second order
   C. lever of the third order
   D. lever of the fourth order

_____ **9.** What wheel brake unit does friction brakes use?
   A. Drum brakes
   B. Jake brakes
   C. Disc brakes
   D. Both A and C
_____ **10.** What type of parking brake uses a small drum brake to prevent the drive shaft from turning?
   A. Drum-style parking brake
   B. Transmission-mounted parking brake
   C. Top hat parking brake
   D. Electric parking brake

True/False

If you believe the statement to be more true than false, write the letter "T" in the space provided. If you believe the statement to be more false than true, write the letter "F".

_____ **1.** The band braking system was used for more than 2,000 years with virtually no change.
_____ **2.** Giving greater control of the braking system to the computer increases driving safety.
_____ **3.** Aggressive driving causes tires to become hot and possibly overheated, thus reducing the tire's ability to obtain maximum traction.
_____ **4.** The service brake is usually operated by hand, but some vehicles use a foot-activated pedal.
_____ **5.** Faster-moving objects have more kinetic energy than slower-moving objects of the same weight.
_____ **6.** An outside force needs to act upon a vehicle to cause it to decelerate, that force comes from the mass of the Earth.
_____ **7.** Static friction is resistance between moving surfaces and is present in standard brakes.
_____ **8.** Most of the heat generated by the braking process radiates into the atmosphere.
_____ **9.** Water fade is caused by water-soaked brake linings acting like a lubricant and lowering the coefficient of friction between the braking surfaces.
_____ **10.** Modern drum and disc brake systems are regularly fitted with an ABS that monitors the speed of each wheel and prevents wheel lock-up or skidding, no matter how hard brakes are applied or how slippery the road surface.

Fill in the Blank

Read each item carefully, and then complete the statement by filling in the missing word(s).

1. A(n) _____-_____-_____ system does away with the hydraulic portion of the brake system and replaces it with sensors, wires, an electronic control unit, and electrically actuated motors to apply individual brake units at each wheel.

2. In a(n) _____ braking system the amount of stopping power is controlled by how much electricity is being generated.

3. There are two brake systems on all vehicles—a _____ brake and a _____ brake.

4. By using _____ of different sizes, hydraulic forces can be increased or reduced, allowing designers to obtain the desired braking force for each wheel.

5. Heavier objects have more _____ energy than lighter objects moving at the same speed.

6. In an automobile, _____, or an increase in kinetic energy, is caused by the power from the engine.

7. The energy used to cause a vehicle to accelerate and decelerate must be _____ from one form of energy to another.

8. _____ _____ is caused by the buildup of heat in the braking surfaces, which get so hot they cannot create any additional heat, leading to a loss of friction.

9. _____ fade is caused by the brake fluid becoming so hot that it boils.

10. Brake systems use _____ and mechanical advantage to apply service and parking brakes.

11. Some vehicles come equipped with _____ pedal assemblies that allow the driver to raise or lower the brake and throttle pedal assembly for personal comfort.

12. On drum brakes, a drum-style parking brake _____ applies the brake shoes against the drum.

Labeling

Label the following diagrams with the correct terms.

1. Weight transfer during braking:

A. _____

B. _____

C. _____

D. _____

2. Basic types of levers:

A. _____

B. _____

C. _____

3. The hydraulic brake system:

A. _____

B. _____

C. _____

D. _____

E. _____

4. Power brake booster:

A. _____

B. _____

C. _____

D. _____

E. _____

5. Top hat design parking brake:

A. _____

B. _____

C. _____

D. _____

6. Drum-style design parking brake:

A. _____

B. _____

C. _____

D. _____

Crossword Puzzle

Use the clues to complete the puzzle.

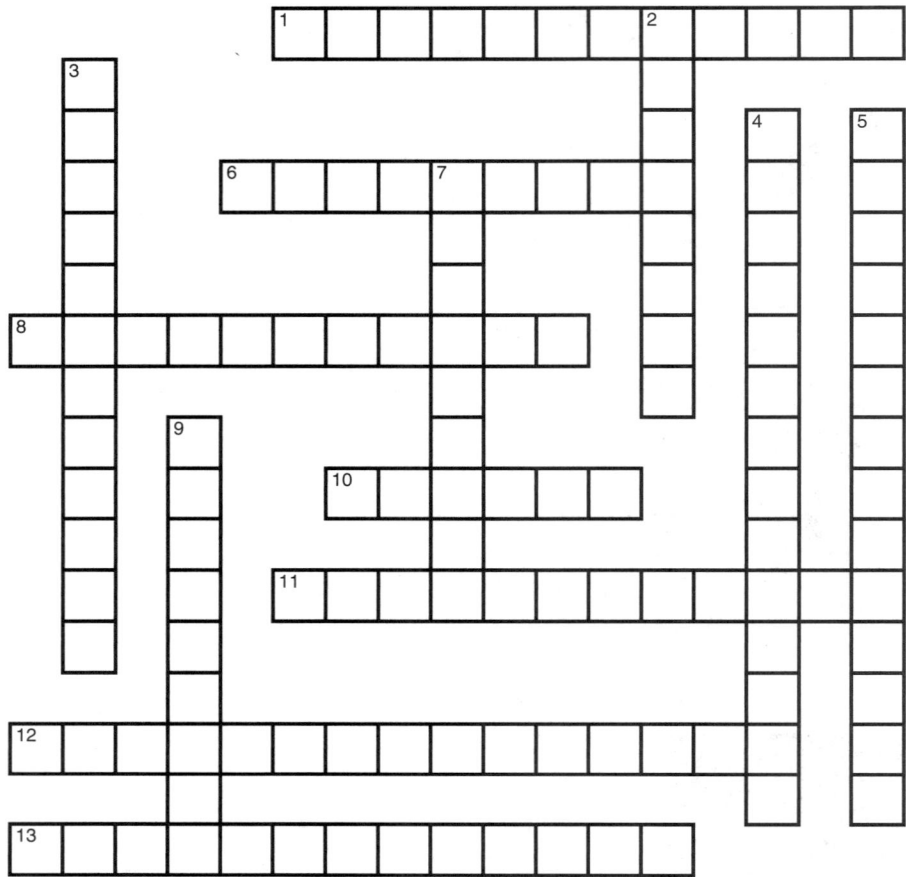

Across

1. An increase in a vehicle's speed.
6. A braking system that uses a metal band lined with friction material to clamp around the outside of a wheel or drum.
8. A brake system that uses leverage to force a friction block against one or more wheels.
10. A system made up of hydraulic and mechanical components designed to slow or stop a vehicle.
11. A decrease in a vehicle's speed.
12. The force created by the rotating wheel when the brakes are applied; it causes the brake components to twist the brake support and ultimately the vehicle in the direction of wheel rotation.
13. The energy of an object in motion; it increases by the square of the speed.

Down

2. A mineral with needle-like fibers that can become embedded in lung tissue and cause cancer.
3. A brake system that is operated while the vehicle is moving to slow or stop the vehicle.
4. Converts the brake pedal force into hydraulic pressure, which is then transmitted via brake lines and hoses to one or more pistons at each wheel brake unit.
5. Weight moving from one set of wheels to the other set of wheels during braking, acceleration, or cornering.
7. The reduction in stopping power caused by a change in the brake system such as overheating, water, or overheated brake fluid.
9. Brake fade caused by water-soaked brake linings.

ASE-Type Questions

Read each item carefully, and then select the best response.

1. Tech A says that regenerative braking converts brake heat into electricity. Tech B says that regenerative braking converts electrical energy into braking energy. Who is correct?
   A. Tech A
   B. Tech B
   C. Both A and B
   D. Neither A nor B

2. Tech A says that an ABS allows the front wheels to be steered during a panic stop. Tech B says that ABS sensor inputs control wheel speed during brake events. Who is correct?
   A. Tech A
   B. Tech B
   C. Both A and B
   D. Neither A nor B

3. Tech A says that water-soaked brake shoes will cause brake fade. Tech B says that disc brakes dissipate heat faster than drum brakes. Who is correct?
   A. Tech A
   B. Tech B
   C. Both A and B
   D. Neither A nor B

4. Tech A says that a brake-by-wire system can start to apply brakes before the driver can step on the brake pedal. Tech B says that brake-by-wire systems use heavy cables to transmit the brake pedal force to the wheel brake units. Who is correct?
   A. Tech A
   B. Tech B
   C. Both A and B
   D. Neither A nor B

5. Tech A says that tire pressure does not affect braking. Tech B says that heavy vehicle loads increase stopping distance. Who is correct?
   A. Tech A
   B. Tech B
   C. Both A and B
   D. Neither A nor B

6. Tech A says that light-duty service brakes are applied hydraulically. Tech B says that light-duty parking brakes are applied hydraulically. Who is correct?
   A. Tech A
   B. Tech B
   C. Both A and B
   D. Neither A nor B

7. Tech A says that the heavier the vehicle, the more stopping power is needed. Tech B says that the faster a vehicle is moving, the more braking power is needed. Who is correct?
   A. Tech A
   B. Tech B
   C. Both A and B
   D. Neither A nor B

8. Tech A says that kinetic energy is created during braking to stop the vehicle. Tech B says that kinetic energy is converted to heat energy during braking. Who is correct?
   A. Tech A
   B. Tech B
   C. Both A and B
   D. Neither A nor B

_____ **9.** Tech A says that the brake pedal uses leverage to multiply foot pressure. Tech B says that when braking hard while moving forward, the vehicle's weight transfers to the rear wheels, increasing their tracion. Who is correct?
  A. Tech A
  B. Tech B
  C. Both A and B
  D. Neither A nor B

_____ **10.** Tech A says that friction brakes can fade due to overheating of the brake lining. Tech B says that friction brakes can fade due to overheating of the brake fluid. Who is correct?
  A. Tech A
  B. Tech B
  C. Both A and B
  D. Neither A nor B

Hydraulic and Power Brakes

Chapter Review

The following activities have been designed to help you refresh your knowledge of this chapter. Your instructor may require you to complete some or all of these activities as a regular part of your training program. You are encouraged to complete any activity that your instructor does not assign as a way to enhance your learning.

Matching

Match the following terms with the correct description or example.

A. Bleeding
B. Brake fluid
C. Brake hose
D. Brake lines
E. Compensating port
F. Free play
G. Inlet port
H. Input force
I. Load transfer
J. Metering valve

K. Output force
L. Outlet port
M. Poppet valve
N. Primary cup
O. Primary piston
P. Quick take-up valve
Q. Residual pressure valve
R. Secondary cup
S. Secondary piston
T. Working pressure

_____ 1. A brake piston in the master cylinder moved directly by the pushrod or the power booster; it generates hydraulic pressure to move the secondary piston.

_____ 2. The amount of clearance between the brake pedal linkage and the master cylinder piston.

_____ 3. Force that equals the working pressure multiplied by the surface area of the output piston, expressed as pounds, newtons, or kilograms.

_____ 4. In drum brake systems, a valve that maintains pressure in the wheel cylinders slightly above atmospheric pressure so air does not enter the system through the seals in the wheel cylinders.

_____ 5. The process of removing air from a hydraulic braking system.

_____ 6. The force applied to the input piston, measured in either pounds or kilograms.

_____ 7. A valve used on vehicles equipped with older rear drum/front disc brakes to delay application of the front disc brakes until the rear drum brakes are applied. Located in line with the front disc brakes.

_____ 8. Hydraulic fluid that transfers forces under pressure through the hydraulic lines to the wheel braking units.

_____ 9. Links the cylinder to the brake lines.

_____ 10. Made of seamless, double-walled steel, and able to transmit over 1,000 psi (6895 kPa) of hydraulic pressure through the hydraulic brake system.

_____ 11. A piston that is moved by hydraulic pressure generated by the primary piston in the master cylinder.

_____ 12. Connects the reservoir with the space around the piston and between the piston cups in a brake master cylinder.

_____ 13. The pressure within a hydraulic system while the system is being operated.

_____ 14. A valve used to release excess pressure from the larger piston in a quick take-up master cylinder once the brake pads have contacted the brake rotors.

_____ 15. A valve that controls the flow of brake fluid at usually preset pressures.

_____ **16.** Connects the brake fluid reservoir to the master cylinder bore when the piston is fully retracted, allowing for expansion and contraction of the brake fluid.

_____ **17.** A seal that prevents loss of fluid from the rear of each piston in the master cylinder.

_____ **18.** A seal that holds pressure in the master cylinder when force is applied to the piston.

_____ **19.** A flexible section of the brake lines between the body and suspension to allow for steering and suspension movement.

_____ **20.** Weight transfer from one set of wheels to the other set of wheels during braking, acceleration, or cornering.

Multiple Choice

Read each item carefully, and then select the best response.

_____ **1.** What law states that pressure applied to a fluid in one part of a closed system will be transmitted without loss to all other areas of the system?
   A. Newton's law
   B. Pascal's law
   C. Kirchhoff's law
   D. Ohm's law

_____ **2.** All of the following are variables to consider when talking about pressure and force in hydraulic systems, _except_:
   A. input force
   B. fluid type
   C. output force
   D. working pressure

_____ **3.** Standard brake fluid is _____, which means it absorbs water.
   A. hydrotropic
   B. aerated
   C. hygroscopic
   D. hydraulic

_____ **4.** Brake fluids are tested to ensure they meet the standards of quality for _____.
   A. stability
   B. resistance to oxidation
   C. boiling point
   D. all of the above

_____ **5.** What type of brake fluid is silicone-based?
   A. DOT 2
   B. DOT 3
   C. DOT 4
   D. DOT 5

_____ **6.** When the brake pedal is released quickly, the use of small holes drilled in the piston so that brake fluid from the reservoir can pass through the inlet port and past the edge of the primary cup, thus preventing a vacuum from being created is called _____.
   A. recuperation
   B. compensating
   C. aeration
   D. residual pressure release

_____ **7.** The _____ master cylinder is a tandem master cylinder used in divided systems.
   A. single-piston
   B. dual-piston
   C. ABS
   D. quick take-up

_____ **8.** What type of flare is sometimes called a bubble flare?
   A. Inverted double flare
   B. International Standards Organization flare
   C. Mushroom flare
   D. DIN flare

_____ **9.** Which of the following components is used to modify pressures within the hydraulic braking system?
   A. Proportioning valves
   B. Pressure differential valves
   C. Antilock hydraulic control units
   D. All of the above

_____ **10.** All of the following conditions will cause the brake warning light to illuminate, *except*:
   A. Parking brake is engaged
   B. Brake fluid is low
   C. Brake pedal is pressed
   D. Prove out circuit

True/False

If you believe the statement to be more true than false, write the letter "T" in the space provided. If you believe the statement to be more false than true, write the letter "F".

_____ **1.** In a closed system, hydraulic pressure is transmitted equally in all directions throughout the system.

_____ **2.** Because silicone-based fluid tends to aerate when forced at high pressure through small passages, it is not to be used in any vehicle equipped with antilock brakes.

_____ **3.** The outlet port adjusts for changes in the volume of the brake fluid ahead of the piston.

_____ **4.** Tandem systems can be split diagonally so one front wheel is paired with the rear wheel on the opposite side of the vehicle in one brake circuit, and vice versa in the other circuit.

_____ **5.** Master cylinder reservoirs can be built into the master cylinder housing or can be a separate unit.

_____ **6.** Master cylinder reservoirs should be filled all the way to the top to prevent air bubbles from entering the system.

_____ **7.** A front-engine rear-wheel drive car has around 40% of its load on its rear wheels and 60% on its front wheels.

_____ **8.** While brake hoses are designed to be flexible, they should never be pinched, kinked, or bent tighter than a specified radius.

_____ **9.** During heavy braking, master cylinder pressure can reach the poppet valve's crack point.

_____ **10.** Adjustable proportioning valves are not recommended for most applications due to the amount of trial and error necessary to set them properly.

_____ **11.** The combination valve can combine the pressure differential valve, metering valve, and proportioning valve(s) in one unit.

_____ **12.** The brake warning light is activated by a normally closed switch located on the brake pedal assembly.

_____ **13.** The vacuum-assisted power booster uses the difference between engine vacuum and atmospheric pressure to increase the force that acts on the master cylinder pistons.

_____ **14.** When the brake pedal is released, the atmospheric valve is opened and the vacuum valve closes.

_____ **15.** While hydraulic brake boosters use power steering fluid to operate the booster, the master cylinder portion of the system still uses brake fluid.

Fill in the Blank

Read each item carefully, and then complete the statement by filling in the missing word(s).

1. The same _____ applied over different-sized surface areas will produce different levels of force.

2. If brake fluid boils, it turns from a liquid to a _____, which is compressible.

3. Brake fluids are graded against compliance standards set by the United States _____ _____ _____.

4. _____-_____ master cylinders have one piston with two cups: a primary cup and a secondary cup.

5. _____ master cylinders combine two master cylinders within a common housing that share a common cylinder bore.

6. Tandem systems can be split _____ to _____ so the front brakes operate from one circuit and the rear brakes from the other.

7. _____ _____-_____ master cylinders are used on disc brake systems that are equipped with low-drag brake calipers.

8. The _____ _____ uses leverage to multiply the effort from the driver's foot to the master cylinder.

9. An alternative to the diagonal, or X, pattern braking system split used in front-engine, front-wheel drive vehicles is an _____ _____.

10. The _____ _____ flare is created by first flaring the end of the tube outward in a Y shape. Then about half of the flared end is folded inside of itself, leaving a double-thick section of brake line on the flared portion of the Y.

11. Many brake hoses use _____ fittings to connect the hose to the wheel unit.

12. _____ valves reduce brake pressure to the rear wheels when their load is reduced during moderate to severe braking.

13. A _____ _____ valve monitors any pressure difference between the two separate hydraulic brake circuits.

14. During hydraulic braking system bleeding, the pressure differential valve may need to be _____.

15. Although not as common as a conventional brake system fitted with a vacuum booster, many vehicles are now equipped with _____ assisted brake boosters.

Labeling

Label the following diagrams with the correct terms.

1. Single piston master cylinder with primary and secondary pumps:

A. _____

B. _____

C. _____

D. _____

E. _____

F. _____

G. _____

H. _____

2. Single piston master cylinder with small holes in the piston to allow for recuperation:

A. _____

B. _____

C. _____

D. _____

E. _____

F. _____

3. Divided hydraulic systems:

A. _____

B. _____

C. _____

4. Flexible brake hose construction:

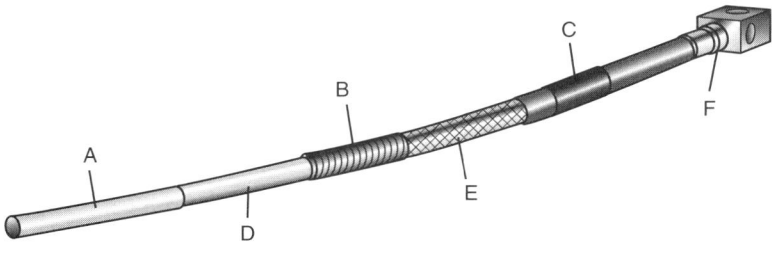

A. _____

B. _____

C. _____

D. _____

E. _____

F. _____

5. Pressure differential valve with a leak in the hydraulic braking system:

A. _____

B. _____

C. _____

D. _____

E. _____

F. _____

G. _____

H. _____

I. _____

J. _____

K. _____

L. _____

M. _____

6. Combination valve:

A. _____

B. _____

C. _____

D. _____

E. _____

F. _____

G. _____

H. _____

I. _____

Skill Drills

Test your knowledge of skill drills by filling in the correct words in the photo captions.

   1. Selecting, Handling, Storing, and Filling Brake Fluid:

Step 1: _____ the specified type of brake fluid in the appropriate service information. Wipe around the master cylinder reservoir cover to prevent any _____ from entering the system. _____ the reservoir cover.

Step 2: Check the fluid level in the reservoir. The fluid should be near the full mark on the _____ of the cylinder or within half an _____ of the top of each _____ if there are no marks.

Step 3: Once you are sure there are no unresolved issues, add the manufacturer's recommended _____ _____ to bring the level to the _____ mark. Replace the cover, and check that it is properly _____. Check for any leaks around the master cylinder. _____ any brake fluid that may have been spilled with fresh clean water.

2. Performing Pressure Bleeding:

Step 1: Prepare the _____ _____, and install it on the vehicle.

Step 2: Install a clear _____ on the farthest bleeder screw, and open it one-quarter to one-half turn. Observe any old brake fluid and _____ _____ coming out.

Step 3: Close off the _____ _____ when the brake fluid is clear and has no bubbles. Tighten it to the manufacturer's specifications. Repeat this procedure for each of the _____ _____ _____, moving closer to the master cylinder, one wheel at a time.

3. Testing Pedal Free Travel and Checking Power Assist Operation:

Step 1: To test brake pedal free travel, with the engine off, _____the brake pedal several times to remove any _____ or hydraulic pressure from the _____ _____. Measure the distance of the brake pedal _____ _____ by depressing the brake pedal by hand until you just feel all of the slack taken up.

Step 2: Measure the distance the _____ _____ travels from the highest reading to the point that all slack is taken up; compare the findings to the specifications. Check _____ _____ operation by beginning with the vehicle engine off. Apply and release the brake pedal _____ or _____ times to bleed off any vacuum or hydraulic pressure in the power booster.

Step 3: Hold the brake pedal down with moderately firm pressure (20–30 lb [9.1–13.6 kg]). Start the engine, and _____ the brake pedal. On vacuum-assisted vehicles, if the pedal drops an _____ or two, the booster is providing boost. If it does not drop, the booster is not providing boost and the following tests will need to be performed to determine the cause of the fault. On some _____-_____ vehicles, when starting the engine with your _____ on the brake pedal, the pedal should rise or fall an inch or so (depending on the vehicle you are working on) if the booster is providing _____.

4. Checking Vacuum Supply to Vacuum-Type Power Booster:

Step 1: With the engine off, remove the _____ _____ from the _____-type booster.

Step 2: Connect a _____ _____ to the vacuum supply end of the _____.

Step 3: Start the engine, and read the vacuum supply _____ to the vacuum-type booster. Vacuum should be _____ than 16 in. Hg (406 mm Hg) on most vehicles. If the reading is insufficient, check for vacuum _____ or _____ in the supply hose or for an improperly _____ engine.

5. Inspecting Brake Lines, Brake Hoses, and Associated Hardware:

Step 1: Safely raise the vehicle on a _____. Trace all brake lines from the master cylinder to each wheel's _____ _____. Inspect the steel brake lines for _____, _____, _____, _____ and _____.

Step 2: Inspect all flexible brake hoses for _____, _____, and _____.

Step 3: _____ any loose _____ and supports.

Crossword Puzzle

Use the clues to complete the puzzle.

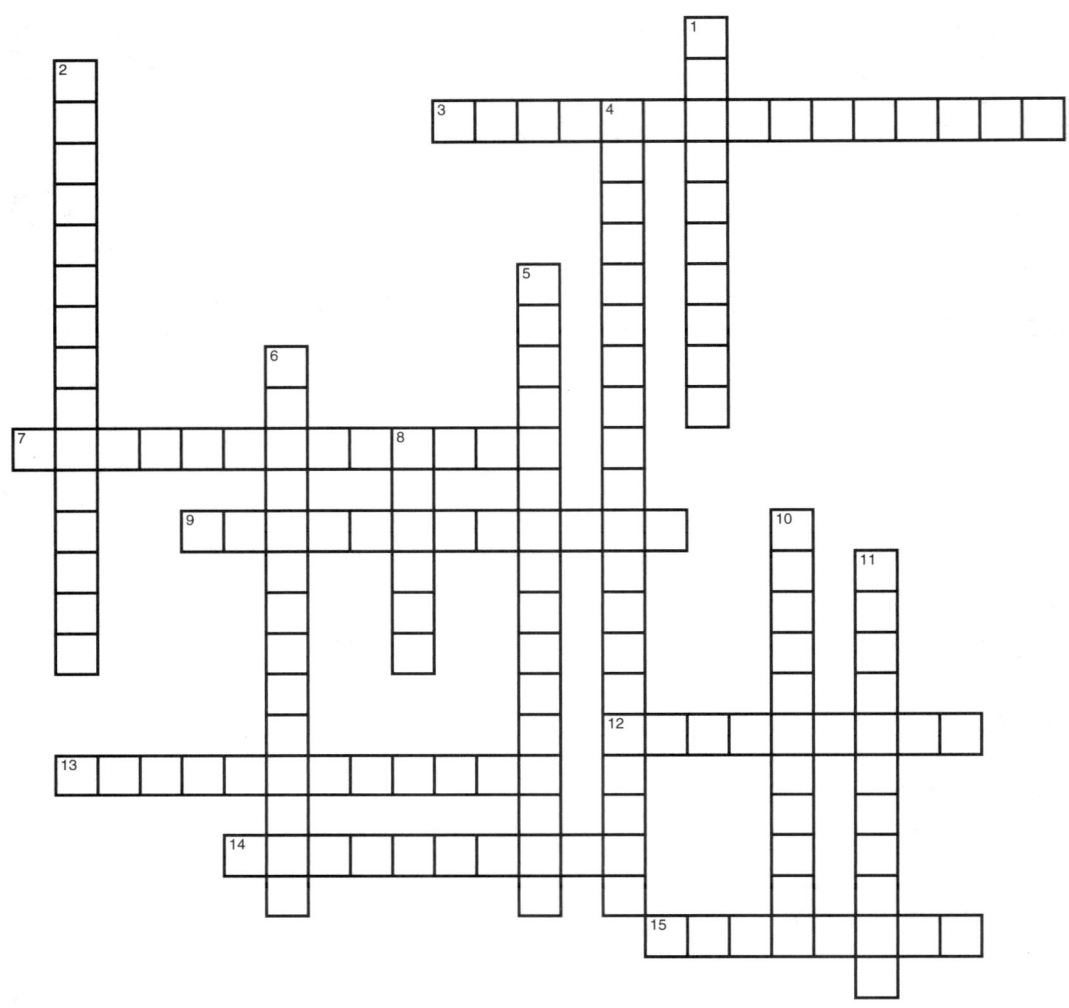

Across

3. A brake system in which the left front wheel is hydraulically paired with the right rear wheel and the right front wheel is paired with the left rear. This preserves 50% of the braking capability if one of the brake circuits begins to leak.

7. A valve used on vehicles equipped with older rear drum/front disc brakes to delay application.

9. Weight transfer from one set of wheels to the other set of wheels during braking, acceleration, or cornering.

12. Connects the reservoir with the space around the piston and between the piston cups in a brake master cylinder.

13. A seal that prevents loss of fluid from the rear of each piston in the master cylinder.

14. The force applied to the input piston, measured in either pounds or kilograms.

15. The process of removing air from a hydraulic braking system.

Down

1. A seal that holds pressure in the master cylinder when force is applied to the piston.

2. The pressure within a hydraulic system while the system is being operated.

4. A master cylinder that has two pistons that operate separate braking circuits so if a leak develops in one circuit, the other circuit can still operate.

5. Connects the brake fluid reservoir to the master cylinder bore when the piston is fully retracted, allowing for expansion and contraction of the brake fluid.

6. A bleeding method where one person manually operates the brake pedal while the other person opens and closes the bleeder screws on the wheel brake units to allow the air and old brake fluid to be pushed out.

8. The tendency to create air bubbles in a fluid.

10. A valve that controls the flow of brake fluid at usually preset pressures.

11. A substance that attracts and absorbs water (e.g., brake fluid).

ASE-Type Questions

Read each item carefully, and then select the best response.

_____ 1. Tech A says that hydraulic pressure is applied equally in all directions throughout a closed system. Tech B says that air in the hydraulic system will cause the brake pedal to be spongy. Who is correct?
   A. Tech A
   B. Tech B
   C. Both A and B
   D. Neither A nor B

_____ 2. Tech A says that a brake pedal that does not return all the way will cause the brake warning light on the instrument panel to stay on. Tech B says that when bleeding brakes, you should normally start at the wheel that is closest to the master cylinder. Who is correct?
   A. Tech A
   B. Tech B
   C. Both A and B
   D. Neither A nor B

_____ 3. Tech A says that brake fluid should periodically be checked for excessive moisture content. Tech B says that brake fluid should be replaced every 12,000 miles. Who is correct?
   A. Tech A
   B. Tech B
   C. Both A and B
   D. Neither A nor B

_____ 4. Tech A says that DOT 5 brake fluid should be used in all vehicles today because it is silicone based and will not absorb water. Tech B says that mixing DOT 4 and DOT 3 is not recommended because of different boiling points, but both are hydroscopic. Who is correct?
   A. Tech A
   B. Tech B
   C. Both A and B
   D. Neither A nor B

_____ 5. Tech A says that the secondary piston in a master cylinder is operated by hydraulic force. Tech B says that the brake pedal return spring returns the master cylinder pistons to their original position. Who is correct?
   A. Tech A
   B. Tech B
   C. Both A and B
   D. Neither A nor B

_____ 6. Tech A says that the low level brake fluid switch on a master cylinder will turn on the brake warning light when the system is low on fluid. Tech B says that the low level switch also monitors the condition of the fluid and will activate the warning light when the brake fluid needs to be replaced. Who is correct?
   A. Tech A
   B. Tech B
   C. Both A and B
   D. Neither A nor B

_____ 7. Tech A says that the metering valve controls pressure to the rear brakes. Tech B says that the proportioning valve controls pressure to the front brakes. Who is correct?
   A. Tech A
   B. Tech B
   C. Both A and B
   D. Neither A nor B

_____ **8.** Tech A says that the vacuum booster uses vacuum and atmospheric pressure to multiply the driver's foot pressure applied to the master cylinder push rod. Tech B says that the vacuum booster increases the vacuum in the brake system. Who is correct?
  A. Tech A
  B. Tech B
  C. Both A and B
  D. Neither A nor B

_____ **9.** Tech A says that bench bleeding a master cylinder will prevent having to bleed air from the brake lines during replacement. Tech B says that bench bleeding the master cylinder makes bleeding the brakes on the vehicle easier. Who is correct?
  A. Tech A
  B. Tech B
  C. Both A and B
  D. Neither A nor B

_____ **10.** Tech A says that a faulty vacuum booster can affect engine operation. Tech B says that steel brake line can be replaced with a copper line, since it is easier to bend into shape. Who is correct?
  A. Tech A
  B. Tech B
  C. Both A and B
  D. Neither A nor B

Disc Brake System

Chapter Review

The following activities have been designed to help you refresh your knowledge of this chapter. Your instructor may require you to complete some or all of these activities as a regular part of your training program. You are encouraged to complete any activity that your instructor does not assign as a way to enhance your learning.

Matching

Match the following terms with the correct description or example.

A. Backing plate
B. Bearing races
C. Bonded linings
D. Brake booster
E. Brake wash station
F. Caliper
G. Dial indicator
H. Drawing-in method
I. Electronic control module
J. Independent rear suspension

K. Lateral runout
L. Low-drag caliper
M. Off-car brake lathe
N. Parallelism
O. Pushrod
P. Riveted linings
Q. Rotor
R. Sliding or floating caliper
S. Steering knuckle
T. Ventilated rotor

_____ **1.** A method for replacing wheel studs that uses the lug nut to draw the wheel stud into the hub or flange.

_____ **2.** Also called warpage, the side-to-side movement of the rotor surfaces as the rotor turns.

_____ **3.** A piece of equipment designed to safely clean brake dust from drum and disc brake components.

_____ **4.** A type of brake caliper that only has piston(s) on the inboard side of the rotor. The caliper is free to slide or float, thus pulling the outboard brake pad into the rotor when braking force is applied.

_____ **5.** A caliper designed to maintain a larger brake pad-to-rotor clearance by retracting the pistons farther than normal.

_____ **6.** Brake linings riveted to the brake pad backing plate with metal rivets and used on heavier-duty or high-performance vehicles.

_____ **7.** Brake linings that are essentially glued to the brake pad backing plate; more common on light-duty vehicles.

_____ **8.** A type of brake rotor with passages between the rotor surfaces that are used to improve heat transfer to the atmosphere.

_____ **9.** A hydraulic device that uses pressure from the master cylinder to apply the brake pads against the rotor.

_____ **10.** A vacuum or hydraulically operated device that increases the driver's braking effort.

_____ **11.** A type of suspension system where each rear wheel is capable of moving independently of the other.

_____ **12.** A metal plate to which the brake lining is fixed.

_____ **13.** A device that connects the front wheel to the suspension; it pivots on the top and bottom, thus allowing the front wheels to turn.

_____ **14.** A tool used to machine (refinish) drums and rotors after they have been removed from the vehicle.

_____ **15.** Tool used to measure the lateral runout of the rotor.

_____ **16.** A mechanism used to transmit force from the brake pedal to the master cylinder.

_____ **17.** Hardened metal surfaces that roller or ball bearings fit into when a bearing is properly assembled.

_____ **18.** The main rotating part of a disc brake system.

_____ **19.** Also called thickness variation; both surfaces of the rotor should be perfectly parallel to each other so brake pulsations do not occur.

_____ **20.** A computer that receives signals from input sensors, compares that information with preloaded software, and sends an appropriate command signal to output devices; used to manage the antilock brake system.

Multiple Choice

Read each item carefully, and then select the best response.

_____ **1.** In high-performance vehicles, the _____ are made from composite materials, ceramics, or carbon fiber; otherwise, they are usually made of cast iron.
   A. brake pads
   B. calipers
   C. rotors
   D. all of the above

_____ **2.** Disc brake caliper assemblies are bolted to the _____.
   A. axle housing
   B. wheel hub
   C. steering knuckle
   D. either A or C

_____ **3.** Disc brake caliper pistons are sealed by a stationary square section sealing ring, also called a _____.
   A. square cut O-ring
   B. caliper gasket
   C. square-to-round gasket
   D. square oil seal

_____ **4.** Manufacturers have dealt with the corrosion issue by making caliper pistons out of _____, which does not corrode or rust.
   A. aluminum
   B. carbon fiber
   C. phenolic resin
   D. rubber

_____ **5.** The backing plate has _____ that correctly positions the pad in the caliper assembly and helps the backing plate maintain the proper position to the rotor.
   A. rivets
   B. tabs
   C. bolts
   D. lugs

_____ **6.** The amount of friction between two surfaces is expressed as a ratio and is called the _____.
   A. sliding resistance
   B. coefficient of friction
   C. friction ratio
   D. drag factor

_____ **7.** Today, brakes are manufactured from a variety of different materials including all of the following, _except_:
   A. kevlar
   B. semimetallic materials
   C. asbestos
   D. ceramic materials

_____ **8.** A spring steel _____ mounted to the brake pad may be used to notify the driver that the brake pads are worn to their minimum limit.
   A. rivet
   B. scratcher
   C. needle
   D. squealer

_____ **9.** Which type of rotors are less expensive and usually found on smaller vehicles?
- **A.** Solid
- **B.** Ventilated
- **C.** Stainless steel
- **D.** Phenolic resin

_____ **10.** Which type of parking brake is engaged by pushing a button on the dash?
- **A.** Top hat
- **B.** Integrated mechanical
- **C.** Electric
- **D.** Automatic

True/False

If you believe the statement to be more true than false, write the letter "T" in the space provided. If you believe the statement to be more false than true, write the letter "F".

_____ **1.** The hub can be part of the brake rotor or a separate assembly that the rotor slips over and is bolted to by the lug nuts.

_____ **2.** Disc brake pads require much lower application pressures to operate than drum brake shoes because they are self-energizing.

_____ **3.** The sliding or floating caliper has brake pads located on each side of the rotor, but all of the pistons are only on one side, usually the inside of the rotor.

_____ **4.** Bonded brake linings are less susceptible to failure under high temperatures.

_____ **5.** The lower the edge code letter, the less friction the material has, and the harder the brake pedal must be applied to achieve a given amount of stopping power.

_____ **6.** Society of Automotive Engineers standards assure that an EE-rated lining from one manufacturer will have the same braking characteristics as an EE-rated lining from another manufacturer.

_____ **7.** Technicians can apply a high-temperature liquid rubber compound to the back of the brake pad that stays flexible, absorbs brake pad vibrations, and helps reduce brake noise.

_____ **8.** Lateral runout tends to move the caliper pistons in the same direction as each other, so brake fluid is not pushed back to the master cylinder.

_____ **9.** Solid rotors are used to improve heat transfer to the atmosphere.

_____ **10.** Most disc brake rotors are stamped with the manufacturer's minimum thickness specification.

Fill in the Blank

Read each item carefully, and then complete the statement by filling in the missing word(s).

1. The purpose of the _____ _____ system is to provide an effective means to slow the vehicle under a variety of conditions in an acceptable distance and manner.

2. Calipers use hydraulic pressure from the _____ _____ to apply the brake pads.

3. On rear-wheel drive vehicles, the _____ is mounted onto the driving axle or hub and may be held in place by the wheel.

4. A(n) _____ _____ on the top of the piston bore allows for the removal of air within the disc brake system as well as helping with performing routine brake fluid changes.

5. _____-_____ calipers are designed to maintain a larger brake pad-to-rotor clearance by retracting the pistons a little bit farther.

6. Floating calipers are mounted in place by _____ _____ and _____.

7. The Society of Automotive Engineers has adopted letter codes to rate brake lining materials' coefficient of friction. The rating is written on the edge of the friction linings and is called the _____ _____.

8. Adding brake pad _____ and _____ to the brake pads help cushion the brake pad and absorb some of the vibration.

9. Incorporating _____ tangs on the brake pad backing plate allow technicians to crimp the tangs so they are more firmly mounted in the caliper.

10. Most _____ _____ are mechanically applied by use of a cable and ratcheting lever assembly.

Labeling

Label the following diagrams with the correct terms.

1. Disc versus drum brakes:

A. _____

B. _____

C. _____

D. _____

2. The master cylinder converts the pedal force into hydraulic pressure:

A. _____

B. _____

C. _____

D. _____

3. Identify the disc brake tools:

A. _____   B. _____

C. _____

D. _____

E. _____

F. _____

G. _____

H. _____

I. _____

J. _____

Crossword Puzzle

Use the clues in the column to complete the puzzle.

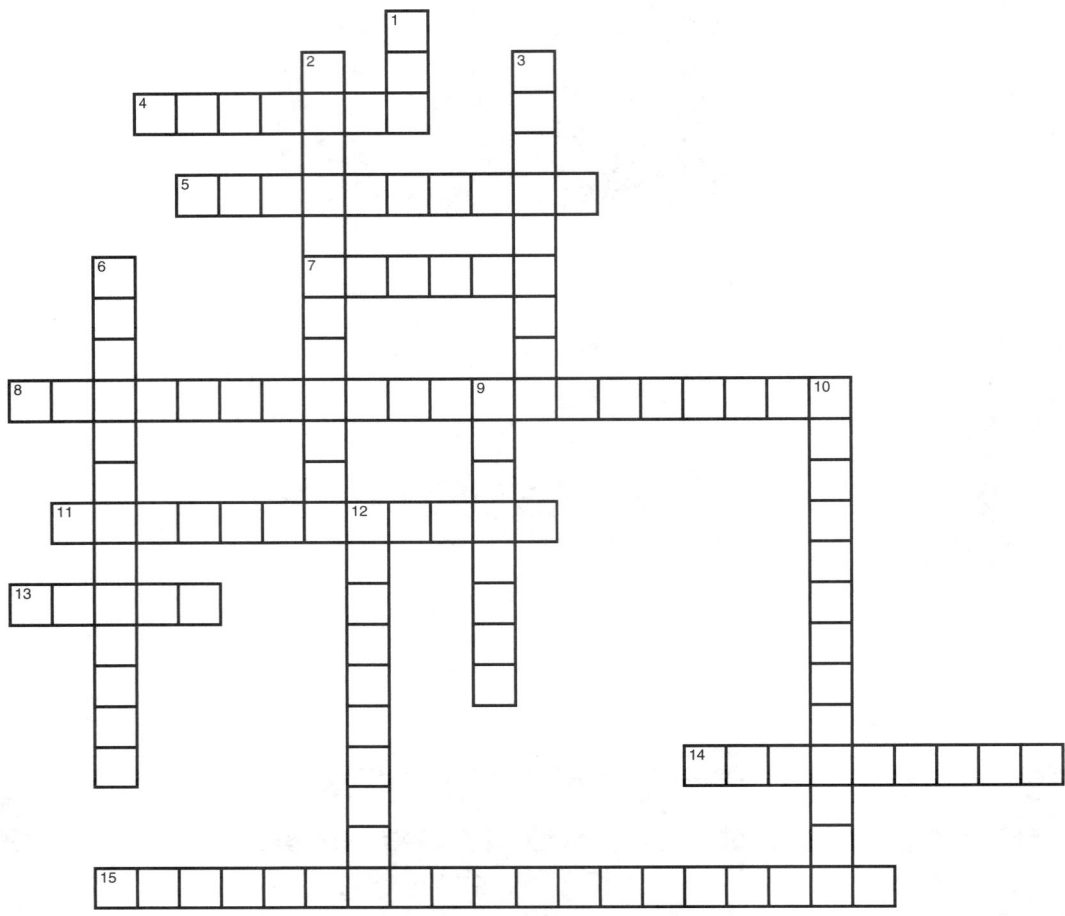

Across

4. An insert with an inner bearing surface that is fitted into a hole in an object, allowing the object to rotate or slide on a pin or shaft.

5. Threaded fasteners that are pressed into the wheel hub flange and used to bolt the wheel onto the vehicle.

7. A tool used to push pistons back into the caliper bore on nonintegrated parking brakes.

8. A method for replacing wheel studs that uses a press to force the wheel stud into the flange until it bottoms out.

11. A screw that allows air and brake fluids to be bled out of a hydraulic brake system when it is loosened and seals the brake fluid in when it is tightened.

13. The main rotating part of a disc brake system.

14. A thin, spring steel wear indicator that is fixed to the backing plate of the brake pad; it emits a high-pitched squeal when the brakes are applied if the brake pads have become too thin.

15. A tool used to grip caliper pistons while removing them.

Down

1. A flange that is shaped to assist with aligning objects on other objects.

2. A type of brake bolted firmly to the steering knuckle or axle housing, having at least one piston on both sides of the rotor.

3. Pins that allow the caliper to move in and out as the brakes operate and as the brake pads wear.

6. Small tabs on the brake pad backing plate that are crimped on to the caliper, creating a secure fit and reducing noise.

9. A code printed on the edge of a friction lining that describes its coefficient of friction.

10. Tool used to measure the lateral runout of the rotor.

12. A type of brake rotor made of solid metal.

ASE-Type Questions

Read each item carefully, and then select the best response.

_____ **1.** Tech A says that disc brakes operate on the principle of friction. Tech B says that disc brakes operate on the principle of regeneration. Who is correct?
   A. Tech A
   B. Tech B
   C. Both A and B
   D. Neither A nor B

_____ **2.** Tech A says that some vehicles use fixed calipers. Tech B says that some vehicles use sliding/floating calipers. Who is correct?
   A. Tech A
   B. Tech B
   C. Both A and B
   D. Neither A nor B

_____ **3.** Tech A says that disc brakes require higher application pressures than drum brakes. Tech B says that disc brakes are self-energizing. Who is correct?
   A. Tech A
   B. Tech B
   C. Both A and B
   D. Neither A nor B

_____ **4.** Tech A says that fixed calipers use one or more pistons only on one side of the rotor. Tech B says that sliding/fixed calipers use one or more pistons on both sides of the rotor. Who is correct?
   A. Tech A
   B. Tech B
   C. Both A and B
   D. Neither A nor B

_____ **5.** Tech A says that calipers use a round section O-ring to seal each piston. Tech B says that calipers use a square section O-ring to seal each piston. Who is correct?
   A. Tech A
   B. Tech B
   C. Both A and B
   D. Neither A nor B

_____ **6.** Tech A says that pistons plated with chrome resist rust. Tech B says that pistons made of phenolic resin do not corrode and rust. Who is correct?
   A. Tech A
   B. Tech B
   C. Both A and B
   D. Neither A nor B

_____ **7.** Tech A says that some vehicles are equipped with a spring-steel brake pad wear indicator that drags on the rotor when the lining thickness is low. Tech B says that some vehicles are equipped with an electric brake pad wear sensor that activates a warning on the dash. Who is correct?
   A. Tech A
   B. Tech B
   C. Both A and B
   D. Neither A nor B

_____ **8.** Tech A says that rotors should be measured for thickness variation (parallelism). Tech B says that rotors should be measured for lateral runout. Who is correct?
   A. Tech A
   B. Tech B
   C. Both A and B
   D. Neither A nor B

_____ **9.** Tech A says that rotors that are too thin cannot handle as much heat and will experience brake fade sooner. Tech B says that brake pedal pulsation is the result of air in the hydraulic system. Who is correct?
   A. Tech A
   B. Tech B
   C. Both A and B
   D. Neither A nor B

_____ **10.** Tech A says that a micrometer is used to measure rotor thickness variation. Tech B says that a micrometer is used to measure rotor lateral runout. Who is correct?
   A. Tech A
   B. Tech B
   C. Both A and B
   D. Neither A nor B

Drum Brake System

Chapter Review

The following activities have been designed to help you refresh your knowledge of this chapter. Your instructor may require you to complete some or all of these activities as a regular part of your training program. You are encouraged to complete any activity that your instructor does not assign as a way to enhance your learning.

Matching

Match the following terms with the correct description or example.

A. Anchor pin
B. Automatic brake self-adjuster
C. Backing plate
D. Brake drum
E. Duo-servo drum brake system
F. Hold-down springs
G. Hold-down spring tool
H. Leading/trailing shoe drum brake system

I. Parking brake mechanism
J. Self-energizing
K. Servo action
L. Specialty springs
M. Springs and clips
N. Twin leading shoe drum brake system
O. Wheel cylinder

_____ 1. Type of brake shoe arrangement where one shoe is positioned in a leading manner and the other shoe in a trailing manner.

_____ 2. A drum brake design where one brake shoe, when activated, applies an increased activating force to the other brake shoe, in proportion to the initial activating force; further enhances the self-energizing feature of some drum brakes.

_____ 3. Springs that hold the brake shoes against the backing plate.

_____ 4. A mechanism that operates the brake shoes or pads to hold the vehicle stationary when the parking brake is applied.

_____ 5. Brake shoe arrangement in which both brake shoes are self-energizing in the forward direction.

_____ 6. A short, wide, hollow cylinder that is capped on one end and bolted to a vehicle's wheel; it has an inner friction surface that the brake shoe is forced against.

_____ 7. A hydraulic cylinder with one or two pistons, seals, dust boots, and a bleeder screw that pushes the brake shoes into contact with the brake drum to slow or stop the vehicle.

_____ 8. Springs used to return links and levers on the parking brake system or the self-adjuster mechanism.

_____ 9. A system on drum brakes that automatically adjusts the brakes to maintain a specified amount of running clearance between the shoes and drum.

_____ 10. A system that uses servo action in both the forward and reverse direction.

_____ 11. Various devices that hold the brake shoes in place or return them to their proper place.

_____ 12. A tool used for removing and installing hold-down springs.

_____ 13. A component of the backing plate that takes all of the braking force from the brake shoes.

_____ 14. A stamped steel plate, bolted to the steering or suspension, which supports the wheel cylinder, brake shoes, and other hardware.

_____ 15. The property of drum brakes that assists the driver in applying the brakes; when brake shoes come into contact with the moving drum, the friction tends to wedge the shoes against the drum, thus increasing the braking force.

Multiple Choice

Read each item carefully, and then select the best response.

_____ 1. All of the following are main components of the drum brake system, *except*:
   A. wheel cylinder
   B. brake shoes
   C. rotor
   D. parking brake mechanism

_____ 2. In drum brake systems, when the brake pedal is depressed, a _____ transfers the force to a hydraulic master cylinder.
   A. linkage
   B. pushrod
   C. cable
   D. brake line

_____ 3. All of the following are types of drum brake systems, *except*:
   A. twin leading shoe
   B. leading/trailing shoe
   C. twin trailing shoe
   D. duo-servo

_____ 4. Brake drums are usually made out of _____ due to its ability to withstand high temperatures, absorb a lot of heat, and maintain its shape.
   A. cast iron
   B. stainless steel
   C. chrome
   D. phenolic resin

_____ 5. The _____ must be able to take all of the braking force when the brakes are applied, so it must be strong and firmly attached to the backing plate.
   A. wheel cylinder
   B. anchor pin
   C. brake drum
   D. primary shoe

_____ 6. A reduction in the coefficient of friction capability as the heat in brake pads and linings builds up is called _____.
   A. bonding fail
   B. slippage
   C. brake fade
   D. wear

_____ 7. Which of the following can cause drum brakes to make a groaning noise?
   A. Excessive brake dust
   B. Overheating
   C. Fluid leak
   D. Worn brake pads

_____ 8. What type of brake springs are generally quite stiff, making them a challenge to install?
   A. Specialty springs
   B. Hold-down springs
   C. Torsion springs
   D. Return springs

_____ 9. What type of brake springs can be of all different shapes and sizes and can be used to push or pull components into their proper position?
   A. Torsion springs
   B. Return springs
   C. Specialty springs
   D. Hold-down springs

_____ **10.** What type of self–adjuster uses two toothed pieces held in contact with each other by spring pressure that can slide over each other in one direction, but hold in the other direction?

    A. Star wheel type

    B. Ratcheting-style

    C. Servo-style

    D. Cable-style

True/False

If you believe the statement to be more true than false, write the letter "T" in the space provided. If you believe the statement to be more false than true, write the letter "F".

_____ **1.** The master cylinder converts brake pedal force into hydraulic pressure.

_____ **2.** Trailing shoes are self-energizing.

_____ **3.** Brake drums provide the rotating friction surface that the brake lining contacts.

_____ **4.** Hubless-style drums have a one-piece integrated hub/drum assembly.

_____ **5.** Cylinder bores on aluminum wheel cylinders are usually honed to help them resist corrosion.

_____ **6.** Drum brakes are usually designed so that the condition of the lining can only be checked once the drum has been removed.

_____ **7.** In a duo-servo brake installation the matching shoes belong on the same side of the vehicle.

_____ **8.** The lining on brake shoes is much thinner than on disc brake pads.

_____ **9.** If you switch a self-adjuster from one side of the vehicle to the other it will retract the adjustment, causing the brake shoe clearance to increase as the brakes adjust.

_____ **10.** Drum parking brake systems mechanically apply the regular service brake shoes.

Fill in the Blank

Read each item carefully, and then complete the statement by filling in the missing word(s).

1. The drum is bolted to the vehicle's axle _____ by the lug nuts.

2. Each drum brake has two brake shoes with a friction material called a(n) _____ attached.

3. Manufacturers might use linings with different _____ _____ _____ for each of the shoes to get the desired braking load between the two shoes.

4. Brake drums are machined to a specific diameter from the manufacturer, which is called its _____ diameter.

5. All of the brake unit components, except the _____ _____, are mounted on a backing plate bolted to the vehicle axle housing or suspension.

6. The cylinder bore, or inside diameter of the cylinder, is created by drawing a properly sized _____ _____ through the bore.

7. The _____ _____ is a hollow screw with a taper on the end that mates with a matching tapered seat in the wheel cylinder.

8. Most modern vehicles use _____-_____ wheel cylinders because they are simpler to design, install, and bleed.

9. The terms primary and secondary refer to the _____ _____ in a duo-servo brake system.

10. Since 1968 manufacturers have incorporated a _____-_____ mechanism into their drum brake systems that is capable of maintaining proper shoe-to-drum clearance.

Labeling

Label the following diagrams with the correct terms.

1. The main components of a drum brake system:

A. _____

B. _____

C. _____

D. _____

E. _____

F. _____

G. _____

H. _____

I. _____

J. _____

K. _____

L. _____

2. The leading/trailing shoe drum brake system:

A. _____

B. _____

C. _____

D. _____

3. The duo-servo drum brake system:

A. _____

B. _____

C. _____

D. _____

E. _____

4. Star wheel assembly:

A. _____

B. _____

C. _____

D. _____

E. _____

5. Parking brake assembly for a drum brake:

A. _____

B. _____

C. _____

D. _____

E. _____

F. _____

6. Drum brake tools:

A. _____

B. _____

C. _____

D. _____

E. _____

F. _____

G.

G. _____

H.

H. _____

I.

I. _____

J.

J. _____

Skill Drills

Test your knowledge of skill drills by filling in the correct words in the photo captions.

1. Refinishing Brake Drums:

Step 1: Research the brake lathe manufacturer's procedure for properly _____ the drum. Clean any nicks, burrs, or debris from the mounting surfaces of the _____. Mount the drum on the brake lathe. Check to see that the drum is running _____ on the lathe.

Step 2: Install the _____-_____ band on the drum.

Step 3: Set the position of the _____ _____ so the brake drum is close to the brake _____ when the cutting bit is in the far corner of the drum. Make sure the cutting bits will not _____ the face of the drum, and move the cutting head about 0.5" (12.7 mm) in from the _____ of the drum. Turn on the brake lathe, and set the _____ of the cutting tool so it just touches the _____ of the drum. Slowly remove the ridge.

Step 4: Run the drum all the way in so the cutting bit is in the _____ _____ of the drum.

Step 5: Set the cutting bit to the proper depth and _____ it in place. Engage the automatic _____ feed, and set it to the proper _____ (if equipped), lock it in place, and watch for proper machining action. Repeat this step until the worn surface areas have been _____ all the way around the surface of the drum. If the brake lathe is not a single-cut machine, perform a _____ _____ on the drum.

Step 6: Move the drum well away from the cutting bit, and use _____ to give the drum surface a nondirectional finish. Remeasure the drum _____, and discard if _____ specifications.

2. Removing, Cleaning, Inspecting, and Reassembling a Duo-Servo Brake:

Step 1: Research the procedure for _____ the brake assembly. Clean brake shoes, hardware, and _____ _____ using equipment and procedures for dealing with _____/dust.

Step 2: To disassemble a _____-_____ brake, first remove the return springs, cable guide (if installed), and _____ guide. Remove the parking brake _____ and spring. Remove the primary shoe _____-_____ _____, retainer, and pin. Remove the self-adjuster spring and _____ _____ assembly and primary shoe.

Step 3: Remove the _____ hold-down spring, retainer, pin, and secondary shoe. Disassemble the _____ _____ lever from the brake shoe and hardware from the backing plate. Finally, _____ and _____ all parts according to the manufacturer's procedure.

Step 4: To reassemble a duo-servo brake, first reassemble the parking brake lever on the _____ _____ and parking brake cable. Reassemble the star wheel assembly, _____ the floating end, and set aside. Also lubricate the _____ on the backing plate. Install both shoes to the backing plate with the hold-down spring assemblies. Install the shoe guide and self-adjuster cable over the _____ _____.

Step 5: Install the cable _____, and return spring in the secondary shoe. Also align the wheel cylinder _____ in the shoe. Position the secondary shoe in place, and use brake spring pliers to install the _____ _____ over the anchor pin. Install the parking brake strut rod onto the secondary shoe, and _____ the primary shoe engaged with the parking brake strut rod. Install the return spring in the primary shoe, and use _____ _____ _____ to stretch the return spring over the anchor pin. Install the self-adjuster link, cable, and spring into position. Install the star wheel between the _____ of the two shoes. Finally, check the _____of all springs, clips, and levers.

3. Disassembling, Cleaning, Inspecting, and Reassembling a Non-Servo Brake:

Step 1: Research the procedure for disassembling the brake assembly. Clean brake shoes, hardware, and _____ _____ using equipment and procedures for dealing with _____ /dust. To disassemble a non-servo brake disassembly, first remove the hold-down springs, _____, and pins. Spread the _____ apart, and remove the parking brake _____ and self- adjuster components.

Step 2: Remove the _____ _____. Disassemble the parking brake lever from the _____ _____ and hardware from the backing plate. Finally, clean and inspect all parts.

Step 3: To reassemble a _____-_____ assembly, first assemble and _____ the self-adjuster/parking brake strut assembly. Lube the backing plate _____. Place one shoe on the backing plate, and _____ the hold-down spring and pin. Place the self-adjuster/parking brake strut on the installed brake shoe.

Step 4: Place the _____ springs on both shoes, and fit the _____ shoe to the backing plate, being sure to line up the wheel cylinder _____, the self-adjuster, and the parking brake mechanism. Install the _____-_____ spring and pin. Check the fit of all springs, clips, and levers.

4. Removing, Inspecting, and Installing Wheel Cylinders:

Step 1: Use a _____ nut or line wrench to unscrew the brake line from the wheel cylinder. Remove the _____ _____ from the backing plate.

Step 2: Peel back the _____ _____, and check for _____ _____ behind them. Determine any necessary actions.

Step 3: _____ the wheel cylinder and _____ each part. Determine any necessary actions.

Step 4: If the cylinder can be reused, _____ it, preferably with new seals and dust boots. If not, _____ it with a new wheel cylinder. Reinstall the _____ _____ by hand. Install and tighten any mounting screws. After that, tighten the brake line with a flare nut or _____ _____.

5. Pre-Adjusting Brakes and Installing Drums and Wheel Bearings:

Step 1: Make sure the brake shoes are _____ on the backing plate.

Step 2: Set the _____-_____ gauge to the drum _____, and lock it in place.

Step 3: Place the pre-adjustment gauge over the _____ of the _____ _____.

Step 4: Adjust the _____ _____ until the centers of the brake shoes just _____ the pre-adjustment gauge.

Step 5: Test install the _____ _____ to verify the drum fits. Adjust the _____ as necessary. Pre-adjust the _____ _____ according to the manufacturer's procedure. If the drum has serviceable wheel _____, repack, install, adjust, and secure them.

Crossword Puzzle

Use the clues to complete the puzzle.

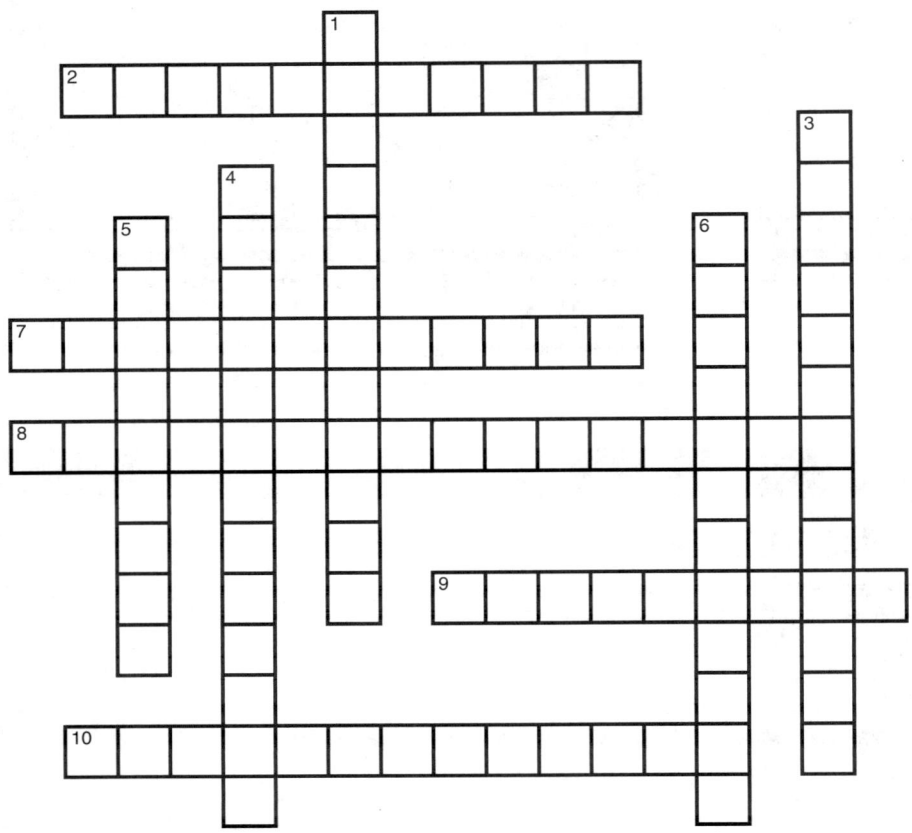

Across

2. A drum brake design where one brake show, when activated, applies an increased activating force to the other brake shoe, in proportion to the initial activating force; further enhances the self-energizing feature of some drum brakes.

7. Brake shoes that are installed so that they are applied in the same direction as the forward rotation of the drum and thus are self-energizing.

8. Springs used to return links and levers on the parking brake system or the self-adjuster mechanism.

9. A component of the backing plate that takes all of the braking force from the brake shoes.

10. A hydraulic cylinder with one or two pistons, seals, dust boots, and a bleeder screw that pushes the brake shoes into contact with the brake drum to slow or stop the vehicle.

Down

1. A stamped steel plate, bolted to the steering or suspension, which supports the wheel cylinder, brake shoes, and other hardware.

3. Springs that retract the brake shoes to their released position.

4. Brake shoes installed so that they are applied in the opposite direction to the forward rotation of the brake drum; not self-energizing and less efficient at developing braking force.

5. A steel show and brake lining friction material that apply force to the brake drum during braking.

6. The inside diameter of a cylinder.

ASE-Type Questions

Read each item carefully, and then select the best response.

_____ **1.** Tech A says that parking brakes on drum brake vehicles use separate brake shoes for backup in an emergency. Tech B says that parking brakes on drum brake vehicles mechanically operate the standard drum brake shoes. Who is correct?
- **A.** Tech A
- **B.** Tech B
- **C.** Both A and B
- **D.** Neither A nor B

_____ **2.** Tech A says that most brake drums are designed to be machined if minor surface issues are present. Tech B says that brake drums can be reused if they are machined over specifications, as long as the surface is smooth. Who is correct?
- **A.** Tech A
- **B.** Tech B
- **C.** Both A and B
- **D.** Neither A nor B

_____ **3.** Tech A says that duo-servo brake shoes are only anchored on the top. Tech B says that typically duo-servo brake shoes adjust automatically during normal brake applications in a forward direction. Who is correct?
- **A.** Tech A
- **B.** Tech B
- **C.** Both A and B
- **D.** Neither A nor B

_____ **4.** Tech A says that when inspecting brake shoes, if the shoes are unequally worn, this could be caused by a stuck wheel cylinder piston. Tech B says that the lining on the primary shoe is typically shorter in length and is installed toward the front of the vehicle. Who is correct?
- **A.** Tech A
- **B.** Tech B
- **C.** Both A and B
- **D.** Neither A nor B

_____ **5.** Tech A says that it is almost impossible to install self-adjusters on the wrong side of the vehicle. Tech B says that grease seals can be reused. Who is correct?
- **A.** Tech A
- **B.** Tech B
- **C.** Both A and B
- **D.** Neither A nor B

_____ **6.** Tech A says to use an air hose to clean the backing plate of dust and contamination. Tech B says to use a brake cleaning solution to clean the backing plate of dust and contamination. Who is correct?
- **A.** Tech A
- **B.** Tech B
- **C.** Both A and B
- **D.** Neither A nor B

_____ **7.** Tech A says that riveted lining is usually for heavy-duty or high-performance vehicles. Tech B says that you need to identify each shoe individually to help ensure proper installation. Who is correct?
- **A.** Tech A
- **B.** Tech B
- **C.** Both A and B
- **D.** Neither A nor B

_____ **8.** Tech A says that a grinding noise in drum brakes generally requires replacing the brake shoes and resurfacing or replacing the drums. Tech B says that a click noise in drum brakes requires replacing the brake shoes and resurfacing or replacing the drums. Who is correct?
- **A.** Tech A
- **B.** Tech B
- **C.** Both A and B
- **D.** Neither A nor B

_____ **9.** Tech A says that brake shoe linings saturated with brake fluid from a leaky wheel cylinder can be cleaned with brake cleaner and reused as long as they aren't worn out. Tech B says that brake shoes should be replaced in axle sets. Who is correct?
- **A.** Tech A
- **B.** Tech B
- **C.** Both A and B
- **D.** Neither A nor B

_____ **10.** Tech A says that when performing a brake job on the rear axle of an older vehicle, inspection finds brake fluid under the dust boot; this suggests wheel cylinder replacement on both rear wheels. Tech B says that when a drum brake return spring has failed, springs on both rear wheels need to be replaced. Who is correct?
- **A.** Tech A
- **B.** Tech B
- **C.** Both A and B
- **D.** Neither A nor B

Wheel Bearings

Chapter Review

The following activities have been designed to help you refresh your knowledge of this chapter. Your instructor may require you to complete some or all of these activities as a regular part of your training program. You are encouraged to complete any activity that your instructor does not assign as a way to enhance your learning.

Matching

Match the following terms with the correct description or example.

A. Anti-friction bearing

B. Ball bearings

C. Bearing packer

D. Castellated nut

E. Cylindrical roller bearing assembly

F. End play

G. Grease seal

H. Interference fit

I. Lithium soap

J. Outer race

K. Preload

L. Running clearance

M. Tapered roller bearing assembly

N. Unitized wheel bearing hub

O. Wheel bearing

_____ 1. A component that is designed to keep grease from leaking out and contaminants from leaking in.

_____ 2. A condition where the wheel bearing components are forced together under pressure and therefore have no end play.

_____ 3. An adjusting nut with slots cut into the top such that it resembles a castle; used with a cotter pin to prevent the nut from turning.

_____ 4. An assembly consisting of the hub, wheel bearing(s), and possibly the wheel flange, which is preassembled and ready to be installed on a vehicle.

_____ 5. The outside component of a wheel bearing that has a smooth, hardened surface for rollers or balls to ride on.

_____ 6. A component that allows the wheels to rotate freely while supporting the weight of the vehicle, made up of an inner race, outer race, rollers or balls, and a cage.

_____ 7. Wheel bearing assemblies that use surfaces that are in rolling contact with each other to greatly reduce friction compared to surfaces in sliding contact.

_____ 8. A condition when two parts are held together by friction because the outside diameter of the inner component is slightly larger than the inside diameter of the outer component.

_____ 9. A tool that forces grease into the spaces between the bearing rollers.

_____ 10. A type of wheel bearing with races and rollers that are tapered in such a manner that all of the tapered angles meet at a common point, which allows them to roll freely and yet control thrust.

_____ 11. A thickening agent for grease to give it the proper consistency.

_____ 12. The in-and-out movement of the hub caused by clearance within the wheel bearing assembly.

_____ 13. The amount of space between wheel bearing components while in operation.

_____ 14. The rolling components of a wheel bearing consisting of hardened balls that roll in matching grooves in the inner and outer races.

_____ 15. A type of wheel bearing with races and rollers that are cylindrical in shape and roll between inner and outer races, which are parallel to each other.

Multiple Choice

Read each item carefully, and then select the best response.

_____ 1. What type of bearings must be serviced periodically by disassembling, cleaning, inspecting, and repacking them with the specified lubricant?
   A. Sealed bearings
   B. Serviceable bearings
   C. Ball bearings
   D. Unitized bearings

_____ 2. What type of wheel bearing assembly is used where heavier loads need to be supported and the wheel bearings are put under a side load condition?
   A. Tapered roller bearing assemblies
   B. Unitized bearings
   C. Cylindrical roller bearing assemblies
   D. Either A or C

_____ 3. Lubrication of serviceable tapered wheel bearing assemblies is usually accomplished by using _____.
   A. wheel bearing grease
   B. lithium soap
   C. gear lube
   D. either A or C

_____ 4. Many seals use a(n) _____ to help hold the lips of the seal in contact with the shaft it is sealing.
   A. O-ring
   B. garter spring
   C. rubber gasket
   D. sealing bushing

_____ 5. In a _____ the weight of the vehicle is fully carried by a pair of tapered roller bearing assemblies, which ride between the hub and axle tube.
   A. semifloating axle
   B. ¾ floating axle
   C. full-floating axle
   D. ½ floating axle

_____ 6. In a(n) _____, the wheel flange is part of the axle, which is supported by a single bearing assembly near the flange end of the axle.
   A. ¾ floating axle
   B. semifloating axle
   C. full-floating axle
   D. solid axle

_____ 7. In a _____ design, there is a single bearing assembly between the outside of the axle tube and the hub.
   A. ¾ floating axle
   B. full floating axle
   C. semifloating axle
   D. splined axle

_____ 8. The level of gear lube should normally be within _____ of the bottom of the fill plug hole.
   A. 0.25"
   B. 0.50"
   C. 0.75"
   D. none of the above

_____ 9. Grease is made out of a base oil, plus a thickening agent such as _____.
   A. lithium soap
   B. polysaccharides
   C. molybdenum thickening agents
   D. either A or C

_____ **10.** On adjustable wheel bearings, the proper clearance must be set using the _____.
   A. keyed washer
   B. adjusting nut
   C. lock cage
   D. lock nut

True/False

If you believe the statement to be more true than false, write the letter "T" in the space provided. If you believe the statement to be more false than true, write the letter "F".

_____ **1.** Sealed bearings are designed so they cannot be disassembled or adjusted.

_____ **2.** The outer bearing assembly is typically larger than the inner bearing assembly.

_____ **3.** Packing is best performed with a bearing packing tool, but it can be successfully performed by hand.

_____ **4.** Roller bearings roll easier than ball bearings since they have a smaller contact area; thus, they provide a small increase in vehicle efficiency.

_____ **5.** Sealed wheel bearings need to be adjusted for proper running clearance after installation.

_____ **6.** On some vehicles equipped with ABS brakes, the ABS sensor is integrated into the unitized wheel bearing assembly.

_____ **7.** Gear lube is somewhat thicker than bearing grease.

_____ **8.** Automotive wheel bearing grease is a thickened lubricant, designated as a plastic solid.

_____ **9.** Sealed bearings and double-row bearing assemblies come from the manufacturer with the proper clearance machined into them.

_____ **10.** On four-wheel drive vehicles, the adjustable wheel bearing locking mechanism commonly includes a keyed washer, adjusting nut, keyed lock washer, and lock nut.

Fill in the Blank

Read each item carefully, and then complete the statement by filling in the missing word(s).

1. In many instances, _____ roller bearing assemblies use the surface of the axle as the inner bearing race.

2. _____ a bearing means that the spaces between the rollers and races are completely filled with grease.

3. Virtually all wheel bearings using a ball bearing assembly are of the _____-_____ ball bearing variety, and they are commonly used in automotive light vehicle applications.

4. In some applications, the _____ _____ is press-fit into the axle housing, which is stationary, and seals against the axle shaft, which is rotating.

5. A(n) _____ _____ is a soft metal pin that can be bent into shape and is used to retain the bearing adjusting nut.

6. In many rear-wheel drive vehicles, the wheel bearing assemblies are open to the axle housing, which is partially filled with _____ _____ that also lubricates the differential assembly.

7. A(n) _____ _____ is usually threaded and can be removed to allow the level of a fluid to be checked and filled.

8. _____ refers to the thickness of the gear lube; the higher the number, the thicker the gear lube.

9. The thickness of grease is graded by the _____ _____ _____ _____.

10. _____ refers to the absence of clearance in the bearing and the specified amount of pressure forcing the bearing components together.

Labeling

Label the following diagrams with the correct terms.

1. Components of a wheel bearing:

A. _____

B. _____

C. _____

D. _____

2. Identify the following components:

A. _____

B. _____

3. Lock nut-style wheel bearing locking mechanism:

A. _____

B. _____

C. _____

D. _____

Crossword Puzzle

Use the clues to complete the puzzle.

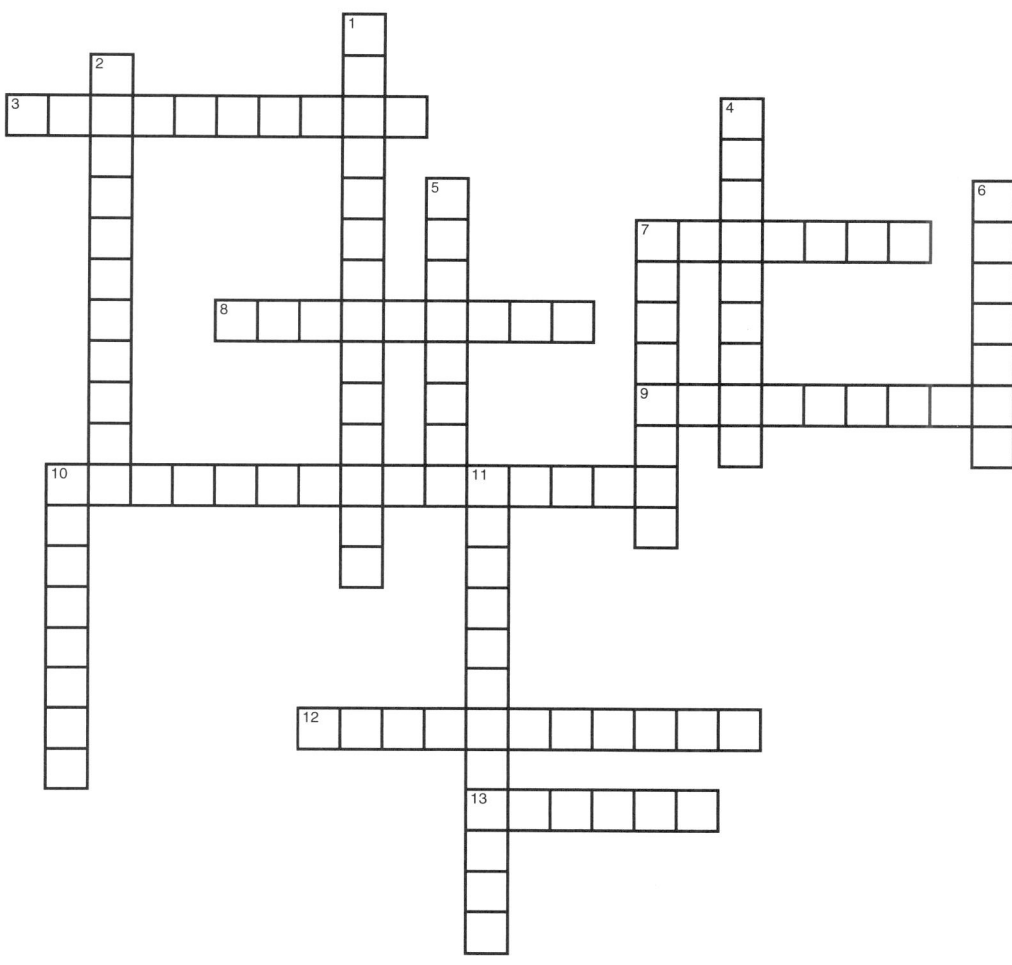

Across

3. A component that is designed to keep grease from leaking out and contaminants from leaking in.

7. The nut that holds the adjusting nut from turning; usually tightened much tighter than the adjusting nut.

8. The inside component of a wheel bearing that has a smooth, hardened surface for rollers or balls to ride on.

9. A one-use soft metal pin that can be bent into shape and is used to retain bearing adjusting nuts.

10. A bearing that uses sliding motion between components, such as a clutch pilot bearing.

12. The component in a wheel bearing that maintains the proper spacing between the roller bearings or ball bearings.

13. A lubricating liquid thickened to make it suitable for use with many wheel bearings.

Down

1. Wheel bearings that are assembled by the manufacturer with the proper lubrication and sealed for life; cannot normally be disassembled.

2. The washer that fits between the adjusting nut and the wheel bearing and that has the center hole keyed to fit a slot on the spindle or axle tube.

4. The measurement of the thickness of a liquid.

5. A type of lubricant primarily used to lubricate transmission and differential gears but also used to lubricate some wheel bearings.

6. A sleeve type bearing made of metal or plastic bearing material used to support a rotating shaft.

7. The stamped sheet metal cap that fits over the bearing adjustment nut and is secured by a cotter pin going through it and the spindle/axle.

10. Usually a threaded plug that can be removed to allow the level of fluid to be checked and filled. This could also be a rubber snap fit plug.

11. The nut used to adjust the end play or preload of a wheel bearing.

ASE-Type Questions

Read each item carefully, and then select the best response.

_____ 1. Tech A says that cylindrical roller bearings can carry more weight than similarly sized ball bearings. Tech B says that tapered roller bearings, used in opposing pairs, control side thrust. Who is correct?
  A. Tech A
  B. Tech B
  C. Both A and B
  D. Neither A nor B

_____ 2. Tech A says that a tapered roller bearing assembly has less rolling resistance than a similarly sized ball bearing assembly. Tech B says that the bearing assembly in a unitized wheel bearing assembly can normally be disassembled, cleaned, and repacked. Who is correct?
  A. Tech A
  B. Tech B
  C. Both A and B
  D. Neither A nor B

_____ 3. Tech A says that over time a grease seal can wear a groove in the sealing surface of the axle or shaft. Tech B says that grease seals need to be replaced every time the bearing is removed. Who is correct?
  A. Tech A
  B. Tech B
  C. Both A and B
  D. Neither A nor B

_____ 4. Tech A says that in a fullfloating axle, the axle does not support the weight of the vehicle. Tech B says that when installing tapered wheel bearings, the final torque should be about 20 ft/lbs. Who is correct?
  A. Tech A
  B. Tech B
  C. Both A and B
  D. Neither A nor B

_____ 5. Tech A says that serviceable wheel bearings can be repacked by removing the dust cap, filling it with grease, and reinstalling it. Tech B says that the cotter pin must be replaced with a new one every time it is removed. Who is correct?
  A. Tech A
  B. Tech B
  C. Both A and B
  D. Neither A nor B

_____ 6. Tech A says that the grease level in the final drive is okay as long as you can touch the level with your finger. Tech B says that the grease level should normally be no more than ¼" below the threads on the fill plug hole. Who is correct?
  A. Tech A
  B. Tech B
  C. Both A and B
  D. Neither A nor B

_____ 7. Tech A says that wheel bearings need to be replaced as a set, bearing and race. Tech B says that the wheel bearings and races on both sides of the vehicle must be replaced if one side fails. Who is correct?
  A. Tech A
  B. Tech B
  C. Both A and B
  D. Neither A nor B

_____ **8.** Tech A says that unitized hubs have a wheel nut with a higher installation torque than serviceable wheel bearings. Tech B says that unitized hubs have the proper bearing end play designed into the assembly once they are torqued properly. Who is correct?
  A. Tech A
  B. Tech B
  C. Both A and B
  D. Neither A nor B

_____ **9.** Tech A says when installing a bearing race you should use a 3-lb hammer and a brass drift. Tech B says that when installing a bearing race you should use a 3-lb hammer and a 3/8" drive extension. Who is correct?
  A. Tech A
  B. Tech B
  C. Both A and B
  D. Neither A nor B

_____ **10.** Tech A says that when a race is fully seated, a sharp metallic sound will be produced when installation is complete. Tech B says that to be sure a race is fully seated, the wheel bearing adjusting nut should be tightened to at least 100 ft-lbs of torque, which will finish seating it. Who is correct?
  A. Tech A
  B. Tech B
  C. Both A and B
  D. Neither A nor B

Electronic Brake Control

Chapter Review

The following activities have been designed to help you refresh your knowledge of this chapter. Your instructor may require you to complete some or all of these activities as a regular part of your training program. You are encouraged to complete any activity that your instructor does not assign as a way to enhance your learning.

Matching

Match the following terms with the correct description or example.

A. Air gap
B. Accumulator
C. Body control module (BCM)
D. Channel
E. Common bore
F. Fault codes
G. High-pressure accumulator
H. Integral ABS system
I. Isolation valve
J. Magneto-resistive sensor

K. Nonintegral ABS systems
L. Oversteer
M. Roll-rate sensor
N. Steering angle sensor
O. Steering wheel position sensor
P. Tone wheel
Q. Understeer
R. Variable reluctance sensor
S. Vehicle speed sensor
T. Wheel speed sensor

_____ **1.** A type of wheel speed sensor that uses the principle of magnetic induction to create its signal.

_____ **2.** A brake system in which the master cylinder, power booster, and HCU are all combined in a common unit.

_____ **3.** A sensor that measures the amount of turning a driver desires. This information is used by the ESC system to know the driver's directional intent.

_____ **4.** An alphanumeric code system used to identify potential problems in a vehicle system.

_____ **5.** A storage container that holds pressurized brake fluid.

_____ **6.** The valve in the HCU that either allows or blocks brake fluid that comes from the master cylinder from entering the HCU hydraulic circuit.

_____ **7.** A brake system in which the master cylinder, power booster, and HCU are all separate units.

_____ **8.** The computer that controls the electrical system in the body of a vehicle, including power windows, door locks, heating and A/C systems, and in some cases the EBC system.

_____ **9.** The component that creates an electrical signal based on the speed of the vehicle, which is sent to the EBCM.

_____ **10.** A sensor that measures the amount of roll around the vehicle's horizontal axis that a vehicle is experiencing.

_____ **11.** A device that creates an analog or digital signal according to the speed of the wheel.

_____ **12.** A storage container designed to contain high-pressure liquids such as brake fluid.

_____ **13.** A sensor that signals to the EBCM both the position and speed of the steering wheel.

_____ **14.** The space or clearance between two components, such as the space between the tone wheel and the pick-up coil in a wheel speed sensor.

_____ **15.** The number of wheel speed sensor circuits and hydraulic circuits the EBCM monitors and controls.

_____ **16.** A condition in which the front wheels are turned further than the direction the vehicle is moving and the front tires are slipping sideways toward the outside of the turn.

_____ **17.** A condition in which the rear wheels are slipping sideways toward the outside of the turn.

_____ **18.** The part of the wheel speed sensor that has ribs and valleys used to create an electrical signal inside of the pick-up assembly.

_____ **19.** When a single cylinder is used for two pistons. A tandem master cylinder would be an example of two pistons in one bore.

_____ **20.** A type of wheel speed sensor that uses an effect similar to a Hall effect sensor to create its signal.

Multiple Choice

Read each item carefully, and then select the best response.

_____ **1.** ABS systems use a computer that sends electrical signals to the _____ that momentarily hold or release hydraulic pressure to that wheel until it speeds up and starts rolling again.
  A. wheel speed sensors
  B. master cylinder
  C. solenoid valves
  D. spool valves

_____ **2.** Which component of the ABS system contains electric solenoid valves controlled by the EBCM to modify hydraulic pressure in each hydraulic circuit?
  A. Master cylinder
  B. Hydraulic control unit
  C. Power booster
  D. Accumulator

_____ **3.** Either separate or combined into a common assembly, the isolation valve and dump valve are controlled by a(n) _____.
  A. electric solenoid
  B. spool valve
  C. check valve
  D. brake pedal sensor

_____ **4.** A _____ system uses separate speed sensors and hydraulic control circuits for each of the four wheels.
  A. single-channel
  B. two-channel
  C. three-channel
  D. four-channel

_____ **5.** If the high-pressure pump fails for any reason, the _____ holds enough brake fluid at high pressure to apply the brakes 10 to 20 times before the boost is used up.
  A. hydraulic control unit
  B. accumulator
  C. master cylinder
  D. reservoir

_____ **6.** A wheel sensor assembly consists of a toothed tone wheel (or tone ring) that rotates with the wheels and a stationary _____ attached to the hub or axle housing.
  A. pick-up assembly
  B. resistor
  C. tooth
  D. relay

_____ **7.** The height of the sine wave is called its _____.
  A. peak
  B. amplitude
  C. phase
  D. duration

_____ **8.** What style of speed sensor is sometimes called a passive system since it is self-contained and needs no outside power to function?
  A. Variable reluctance
  B. Magneto-resistive
  C. Hall effect
  D. Either A or C

_____ **9.** Some ABS systems provide _____ through the ABS warning lamp when a specific terminal is grounded or two specific terminals are shorted together.
   A. blink codes
   B. morse codes
   C. flash codes
   D. either A or C

_____ **10.** An electronic stability control system may integrate a(n) _____ into the basic ABS and TCS systems.
   A. yaw sensor
   B. steering angle sensor
   C. roll-rate sensor
   D. all of the above

True/False

If you believe the statement to be more true than false, write the letter "T" in the space provided. If you believe the statement to be more false than true, write the letter "F".

_____ **1.** Electronic stability control systems take the ABS and TCS systems to the next level by adding sensor information regarding the driver's directional intent.

_____ **2.** Understeer occurs when the vehicle is turning more sharply than the front wheels are being steered.

_____ **3.** Applying the brakes too hard or on a slippery surface can cause the wheels to lock.

_____ **4.** Maximum braking traction occurs when the wheels are rotating 20–25% slower than the vehicle speed.

_____ **5.** Nonintegral ABS systems use a fairly standard tandem master cylinder and a typical vacuum or hydraulic power booster.

_____ **6.** Some hydraulic control units use a single, three-position solenoid valve per circuit, while others use dual, two-position valves per hydraulic circuit.

_____ **7.** The pick-up assembly and tone wheel do not touch each other; a small gap, called an air gap, must be maintained at the specified clearance.

_____ **8.** The magneto-resistive sensor does not function effectively below vehicle speeds of around 5 mph.

_____ **9.** The brake switch is a normally open switch, meaning that if the switch is not affected by any outside force, electrical current will flow through it.

_____ **10.** Many electronic stability control system-equipped vehicles also monitor signals from the throttle position sensor, vehicle speed sensor, and brake pedal position sensor to help prevent a loss of control of the vehicle.

Fill in the Blank

Read each item carefully, and then complete the statement by filling in the missing word(s).

1. In the quest for increased safety, manufacturers developed a series of electronic _____ _____ systems; the first-generation was the antilock brake system.

2. The _____ _____ system applies brake pressure to the slipping tire, which causes more of the engine's torque to be transmitted to the wheel or wheels with the most traction.

3. The _____-_____ _____ system is designed to prevent wheels locking or skidding, no matter how hard the brakes are applied or how slippery the road surface, and to maintain steering control of the vehicle.

4. When the ignition switch is turned on, the ABS controller illuminates the yellow ABS warning lamp and performs an automatic _____-_____ of the system.

5. Drivers need to be taught to expect ABS brake pedal _____ when in a panic stop.

6. The ABS control module (or EBCM) sends commands in the form of _____ _____ to the hydraulic control unit.

7. A(n) _____-_____ system uses one sensor circuit with the speed sensor typically located in the differential and one hydraulic control circuit to control both rear wheels.

8. _____-_____ accumulators hold brake fluid in a spring-loaded chamber when it is released by the dump valves during an EBC event.

9. The electronic brake control module supplies the magneto-resistive or Hall effect sensor systems with a
_____ _____ of between 5 and 12 volts, depending on the manufacturer.

10. On a traction control system _____ _____ direct hydraulic pressure from the
accumulator to the ABS solenoid valves so individual wheel brake units can be applied independently.

Labeling

Label the following diagrams with the correct terms.

1. HCU solenoid valve arrangement:

A. _____

B. _____

C. _____

D. _____

E. _____

F. _____

G. _____

2. Portless ABS master cylinder:

A. _____

B. _____

C. _____

D. _____

E. _____

3. The four types of ABS channels:

A. _____

B. _____

C. _____

D. _____

4. High-pressure accumulator:

A. _____

B. _____

C. _____

D. _____

E. _____

F. _____

G. _____

5. Wheel speed sensor sine wave:

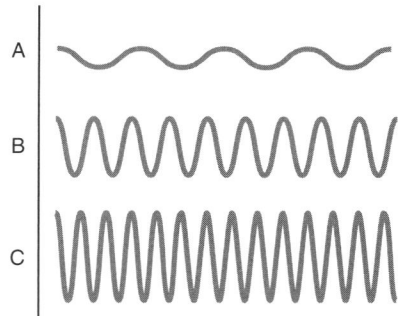

A. _____

B. _____

C. _____

6. An HCU with boost valves:

A. _____

B. _____

C. _____

D. _____

E. _____

F. _____

G. _____

H. _____

Crossword Puzzle

Use the clues to complete the puzzle.

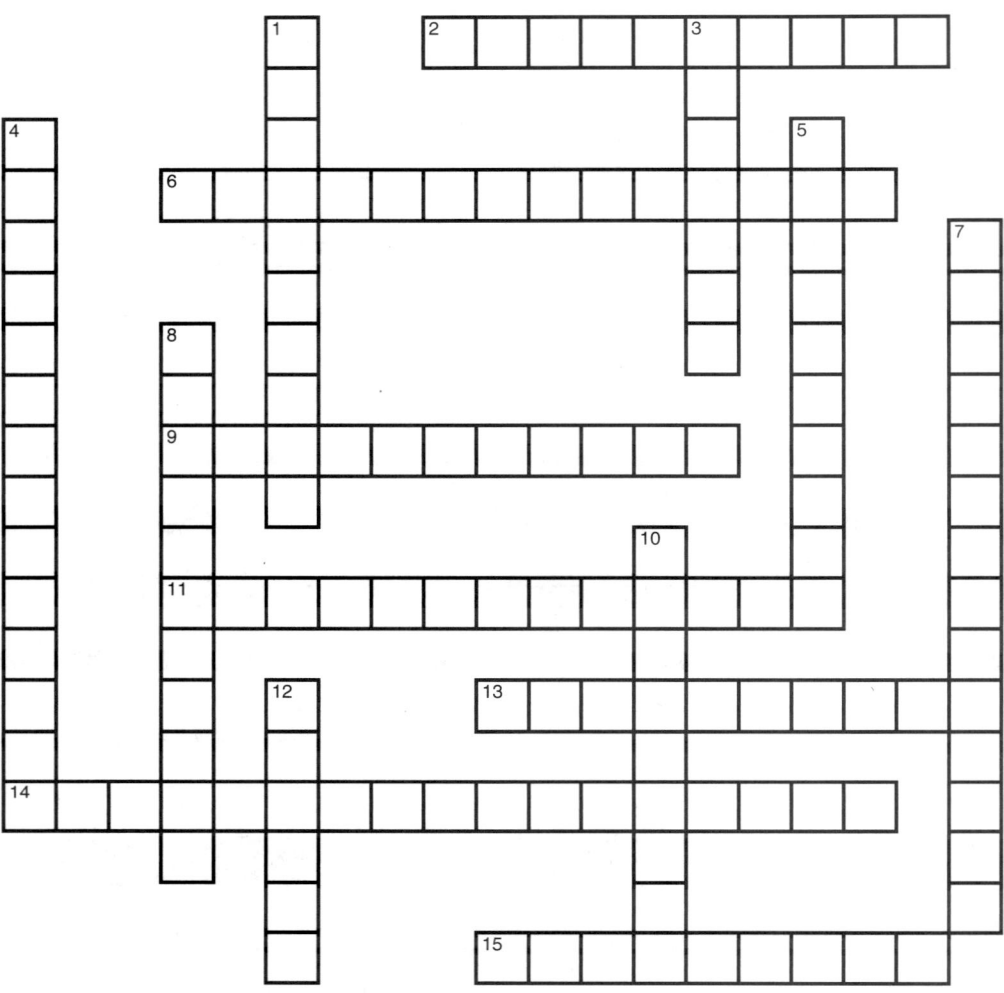

Across

2. An alphanumeric code system used to identify potential problems in a vehicle system.

6. A sensor that measures the amount of roll around the vehicle's horizontal axis that a vehicle is experiencing.

9. A storage container that holds pressurized brake fluid.

11. An electrically operated valve that when used in brake systems is designed to control the flow or brake fluid in the hydraulic system.

13. A valve located in the hydraulic control unit that is controlled by the EBCM; it allows brake fluid under high pressure to flow into the HCU hydraulic circuits to apply the brakes when commanded.

14. An electrically operated valve, which in brake systems is used to control the flow of brake fluid in the hydraulic system.

15. A condition in which the rear wheels are slipping sideways toward the outside of the turn.

Down

1. An electrical effect where electrons tend to flow on one side of a special material when exposed to a magnetic field, causing a difference in voltage across the special material. This effect can be used to determine the position or speed of an object.

3. The number of wheel speed senor circuits and hydraulic circuits the EBCM monitors and controls.

4. The valve in the HCU that either allows or blocks brake fluid that comes from the master cylinder form entering the HCU hydraulic circuit.

5. When a single cylinder is used for two pistons. A tandem master cylinder would be an example of brake fluid in the hydraulic system.

7. A component with a wire coil wrapped around the ferrous metal core; it is used to generate an electrical signal when a magnetic field passes through it.

8. The electrical switch that is activated by the brake pedal; it turn on the brake lights and signals the EBCM that the brakes are being applied.

10. A sensor that measures the amount a vehicle is turning around its vertical axis. This information is used by the ESC system to know how much a vehicle is turning.

12. The space or clearance between two components, such as the space between the tone wheel and the pick-up coil in a wheel speed sensor.

ASE-Type Questions

Read each item carefully, and then select the best response.

_____ **1.** Tech A says that an anti-lock brake system (ABS) helps shorten the stopping distance during a panic stop. Tech B says that antilock brake systems work by increasing the hydraulic pressure in the brake system so the brakes can be applied harder. Who is correct?
   A. Tech A
   B. Tech B
   C. Both A and B
   D. Neither A nor B

_____ **2.** Tech A says that traction control can reduce the power output of the engine to increase traction. Tech B says that electronic stability control increases the risk of rollover. Who is correct?
   A. Tech A
   B. Tech B
   C. Both A and B
   D. Neither A nor B

_____ **3.** Tech A says that an electronic braking system has sensors that monitor wheel speed. Tech B says that under steer is generally easier to recover from than over steer. Who is correct?
   A. Tech A
   B. Tech B
   C. Both A and B
   D. Neither A nor B

_____ **4.** Tech A says that ABS controls braking every time the brakes are used. Tech B says that during an ABS event, it is normal for the brake pedal to pulsate. Who is correct?
   A. Tech A
   B. Tech B
   C. Both A and B
   D. Neither A nor B

_____ **5.** Tech A says that during anti-lock braking, brake fluid may be returned to the master cylinder. Tech B says that solenoid valves in the hydraulic control unit will isolate the master cylinder from the brake circuit when it is in the "hold" mode. Who is correct?
   A. Tech A
   B. Tech B
   C. Both A and B
   D. Neither A nor B

_____ **6.** Tech A says that mismatched tires may cause the ABS system to register a fault code. Tech B says that a TCS may automatically apply brake pressure to a wheel brake unit even if the vehicle is not being braked. Who is correct?
   A. Tech A
   B. Tech B
   C. Both A and B
   D. Neither A nor B

_____ **7.** Tech A says that an ABS key-on system test checks for faults in the vehicle's base brake system. Tech B says that on most vehicles ABS DTCs are stored in memory for later retrieval. Who is correct?
   A. Tech A
   B. Tech B
   C. Both A and B
   D. Neither A nor B

_____ **8.** Tech A says that a yaw sensor tells the computer the vehicle's actual direction. Tech B says that raising a vehicle's curb height has no effect on the ESC system. Who is correct?
   A. Tech A
   B. Tech B
   C. Both A and B
   D. Neither A nor B

_____ **9.** Tech A says that some hydraulic units can boost the brake pressure created by the driver. Tech B says that the dump valve can release pressure in the brake system when ABS is applied during a panic stop. Who is correct?
   A. Tech A
   B. Tech B
   C. Both A and B
   D. Neither A nor B

_____ **10.** Tech A says that some ABS wheel speed sensors create a square wave digital pattern. Tech B says that some wheel speed sensors create an AC sine wave pattern. Who is correct?
   A. Tech A
   B. Tech B
   C. Both A and B
   D. Neither A nor B

Principles of Electrical Systems

Chapter Review

The following activities have been designed to help you refresh your knowledge of this chapter. Your instructor may require you to complete some or all of these activities as a regular part of your training program. You are encouraged to complete any activity that your instructor does not assign as a way to enhance your learning.

Matching

Match the following terms with the correct description or example.

A. Alternating current (AC)
B. Amp
C. Charge carrier
D. Circuit breaker
E. Diode
F. Electrical resistance
G. Energy
H. Ground
I. Insulator
J. Invertor

K. Ohm
L. Phase
M. Polarity
N. Semiconductor
O. Short
P. Sine wave
Q. Solenoid
R. Switch
S. Transformer action
T. Volt

_____ **1.** A term used to describe one set of windings from an alternator or alternating current electric motor.

_____ **2.** A material that has properties that prevent the easy flow of electricity. These materials are made up of atoms with five to eight electrons in the valance ring.

_____ **3.** An electromagnet with a moving iron core that is used to cause mechanical motion.

_____ **4.** A material's property that slows down the flow of electrical current.

_____ **5.** The state of charge, positive or negative.

_____ **6.** A two-lead electronic component that allows current flow in one direction only.

_____ **7.** A mathematical function that describes a repetitive waveform such as an alternating current signal.

_____ **8.** A type of current flow that flows back and forth.

_____ **9.** The transfer of electrical energy from one coil to another through induction in a transformer.

_____ **10.** The ability to do work.

_____ **11.** Also called a short circuit, the flow of current along an unintended route.

_____ **12.** A mobile particle that has a positive or negative charge.

_____ **13.** An electrical device with contacts that turns current flow on and off.

_____ **14.** A device that changes direct current into alternating current.

_____ **15.** An abbreviation for amperes, the unit for current measurement.

_____ **16.** The unit for measuring electrical resistance.

_____ **17.** The unit used to measure potential difference or electrical pressure.

_____ **18.** A material used to make microchips, transistors, and diodes.

_____ **19.** A device that trips and opens a circuit, preventing excessive current flow in a circuit. It is resettable to allow for reuse.

_____ **20.** The return path for electrical current in a vehicle chassis, other metal of the vehicle, or dedicated wire.

Multiple Choice

Read each item carefully, and then select the best response.

_____ 1. Electromotive force is also referred to as _____.
A. voltage
B. resistance
C. amperage
D. electrons

_____ 2. The number of charge carriers, either an excess of electrons or deficiency of electrons can be altered by _____, or adding very small quantities of impurities to a semiconductor material.
A. spiking
B. doping
C. diluting
D. charging

_____ 3. The point where P-type and N-type semiconductors join is called the _____.
A. depletion area
B. hole
C. PN junction
D. carrier

_____ 4. Which type of current flow is produced by a battery?
A. Alternating current
B. Three phase
C. Direct current
D. Sine wave

_____ 5. _____ refers to the number of separate staggered power windings in the motor or alternator.
A. Phase
B. Sine wave
C. Hertz
D. Current

_____ 6. The term _____ describes a low-voltage circuit that does not have a complete circuit and therefore cannot conduct current.
A. closed
B. open
C. short
D. grounded

_____ 7. Which of the following Ohm's law formulas is correct?
A. A = V/R
B. V = A × R
C. R = V ÷ A
D. All of the above

_____ 8. The rate of transforming energy is also known as _____.
A. power
B. voltage
C. work
D. magnetism

_____ 9. In a(n) _____ circuit, all components are connected directly to the voltage supply.
A. series
B. parallel
C. series/ parallel
D. all of the above

_____ 10. Which law states that current entering any junction is equal to the sum of the current flowing out of the junction?
A. Ohm's law
B. Kirchhoff's current law
C. The law of conservation of energy
D. Newton's first law of energy

_____ **11.** A _____ is made up of an electromagnet, a set of switch contacts, terminals, and the case.
  A. thermocouple
  B. solenoid
  C. diode
  D. relay

_____ **12.** An ignition coil can be described as a _____.
  A. solenoid
  B. relay
  C. step-up transformer
  D. step-down transformer

_____ **13.** Mechanical variable resistors with three connections, two fixed and one moveable, are called _____.
  A. thermistors
  B. rheostats
  C. potentiometers
  D. transistor

_____ **14.** All of the following are examples of the type of shielding used in shielded wiring harnesses, except:
  A. Twisted pair
  B. Drain lines
  C. Seamless plastic
  D. Mylar tape

_____ **15.** Which of the following is an example of a crimp type terminal?
  A. Push-on spade
  B. Butt connector
  C. Eye ring
  D. All of the above

True/False

If you believe the statement to be more true than false, write the letter "T" in the space provided. If you believe the statement to be more false than true, write the letter "F".

_____ **1.** A deficiency of electrons gives an atom an overall positive charge.

_____ **2.** Most wiring diagrams are written from the conventional theory perspective, while electronic circuits are typically designed and operate on the electron theory perspective.

_____ **3.** The number of charge carriers, either an excess of electrons or deficiency of electrons (causing holes in a material), can be altered by doping, or adding very large quantities of impurities to a semiconductor material.

_____ **4.** Hertz is the measurement of frequency and indicates the number of cycles per second.

_____ **5.** Voltage drop testing is the best way of finding high resistance in the feed side or ground side of the circuit.

_____ **6.** Since current flowing through a resistance causes heat, you can sometimes locate the high resistance in the circuit just by smelling the wires and connections.

_____ **7.** Ohm's law is a relationship between volts, amps, and ohms, and because they must always balance out, if we know any two of the values, then we can calculate the third.

_____ **8.** A light bulb uses a certain amount of electrical power, but the power used is not an indication of brightness.

_____ **9.** Some capacitors, and most semiconductor components, are polarity sensitive.

_____ **10.** Flasher cans are electronic devices, while flasher controls are mechanical devices.

_____ **11.** A solenoid is an electromechanical device that converts electrical energy into mechanical linear movement.

_____ **12.** In a motor, the armature and brushes act as switches to control the current flow through the windings of the commutator.

_____ **13.** A thermistor is a mechanical variable resistor with two connections.

_____ **14.** Printed circuits are essentially a map of all of the electrical components and their connections.

_____ **15.** Wiring diagrams use abstract graphical symbols to represent electrical circuits and their connection or relationship to other components in the system.

Fill in the Blank

Read each item carefully, and then complete the statement by filling in the missing word(s).

1. _____ _____ are only loosely held by the nucleus and are free to move from one atom to another when an electrical potential (pressure) is applied.

2. Electrical _____, measured in ohms, affects the current flow in a circuit.

3. _____ _____ tells us that if we increase current flow through a resistance, the voltage used by that resistance will increase.

4. A doped semiconductor always has an excess of one type of charge carrier; electrons in excess make it an _____-_____ semiconductor, and holes in excess make it a _____-_____ semiconductor.

5. Within the PN junction some electrons and holes cancel each other out and few charge carriers are present; this very thin _____ _____ acts like an insulator.

6. Voltage can be measured by hooking a voltmeter or _____, set to read voltage across two parts of a circuit where you want to measure the difference in volts.

7. Devices are available to change DC into AC and are called _____.

8. _____ is achieved when an electrical circuit has a continuous and uninterrupted electrical connection and is thereby capable of conducting current and working as designed.

9. In its purest definition, the term _____ describes a circuit fault in which current takes a shorter path, resistance-wise, through an accidental or unintended route.

10. One _____ is produced when 1 volt causes 1 amp of current to flow.

11. The _____ _____ across each resistor can be found by subtracting the voltage after a resistor from the voltage before it, or the difference can be measured.

12. A(n) _____-_____ relay acts like a mechanical relay but does not have any moving parts.

13. The metal core of a solenoid, used by the electromagnet to strengthen the magnetic field, is referred to as a(n) _____.

14. There are two scales used to measure the sizes of wires: the metric wire gauge and the _____ _____ _____.

15. Wiring _____, also known as wiring looms or cable harnesses, are used throughout the vehicle to group two or more wires together within a sheath of either insulating tape or tubing.

Labeling

Label the following diagrams with the correct terms.

1. Parts of an atom (including charge):

A. _____   Charge _____

B. _____   Charge _____

C. _____   Charge _____

2. Ions:

A   B

A. _____

B. _____

3. Insulators:

A   B   C

A. _____

B. _____

C. _____

4. Short circuit:

A. _____

B. _____

C. _____

D. _____

E. _____

F. _____

5. Wiring diagram symbols:

A.

B.

C.

D.

E.

F.

G.

H.

I.

J.

K.

L.

M.

N.

O.

P.

Q.

R.

S.

T.

U.

V.

W.

X.

Y.

Z.

A. _____

B. _____

C. _____

D. _____

E. _____

F. _____

G. _____

H. _____

I. _____

J. _____

K. _____

L. _____

M. _____

N. _____

O. _____

P. _____

Q. _____

R. _____

S. _____

T. _____

U. _____

V. _____

W. _____

X. _____

Y. _____

Z. _____

Skill Drills

Test your knowledge of skill drills by filling in the correct words in the photo captions.

1. Stripping Wire Insulation:

Step 1: Choose the correct _____ _____.

Step 2: Select the hole that matches the _____ of the wire to be stripped. Place the wire in the hole, and close the _____ firmly around it to cut the _____.

Step 3: Remove the insulation. To keep the _____ together, give them a light _____.

2. Installing a Solderless Terminal:

Step 1: Make sure you have the correct size of _____ for the wire to be terminated and the terminal has the correct volt/amp _____. Remove an appropriate amount of the protective _____ from the wire.

Step 2: Lightly _____ the wire _____ and place the terminal onto the _____.

Step 3: Use a proper _____ _____ for the terminal you are crimping. Do not use _____, as they have a tendency to cut through the connection. Select the proper _____.

Step 4: Crimp the _____ section first. Use firm _____ so a good electrical contact will be made, but not _____ force, as this can _____ the pin or terminal.

Step 5: If crimping an _____ terminal, lightly crimp the insulation _____ so they hold the _____ firmly.

3. Soldering Wires and Connectors:

Step 1: Safely position the soldering iron while it is _____ _____. While the soldering iron is heating, _____ an appropriate amount of the protective _____ from the wires with _____ _____.

Step 2: _____ the wires together to make a good _____ _____ between them.

Step 3: _____ the soldering iron tip, and gently heat up the _____ while placing the _____ opposite of the soldering iron. Allow the solder to be _____ into the joint.

Step 4: A good _____ _____ is where the solder has been _____ in.

Step 5: Once the electrical connection has been made and it has _____ enough for you to handle it, slide the insulator _____ cover over the joint and use a _____ _____ to shrink the tubing around the _____.

Step 6: To solder a wire to a terminal _____, it is best to _____ it in place as before and use the solder to "glue" the joint together. Place the heated iron onto the _____ to get it hot enough to _____ the solder applied to the end of the crimped wire tabs. Some solder will be _____ between the terminal and the wire. Cover the terminal with _____-_____ tubing.

Crossword Puzzle

Use the clues to complete the puzzle.

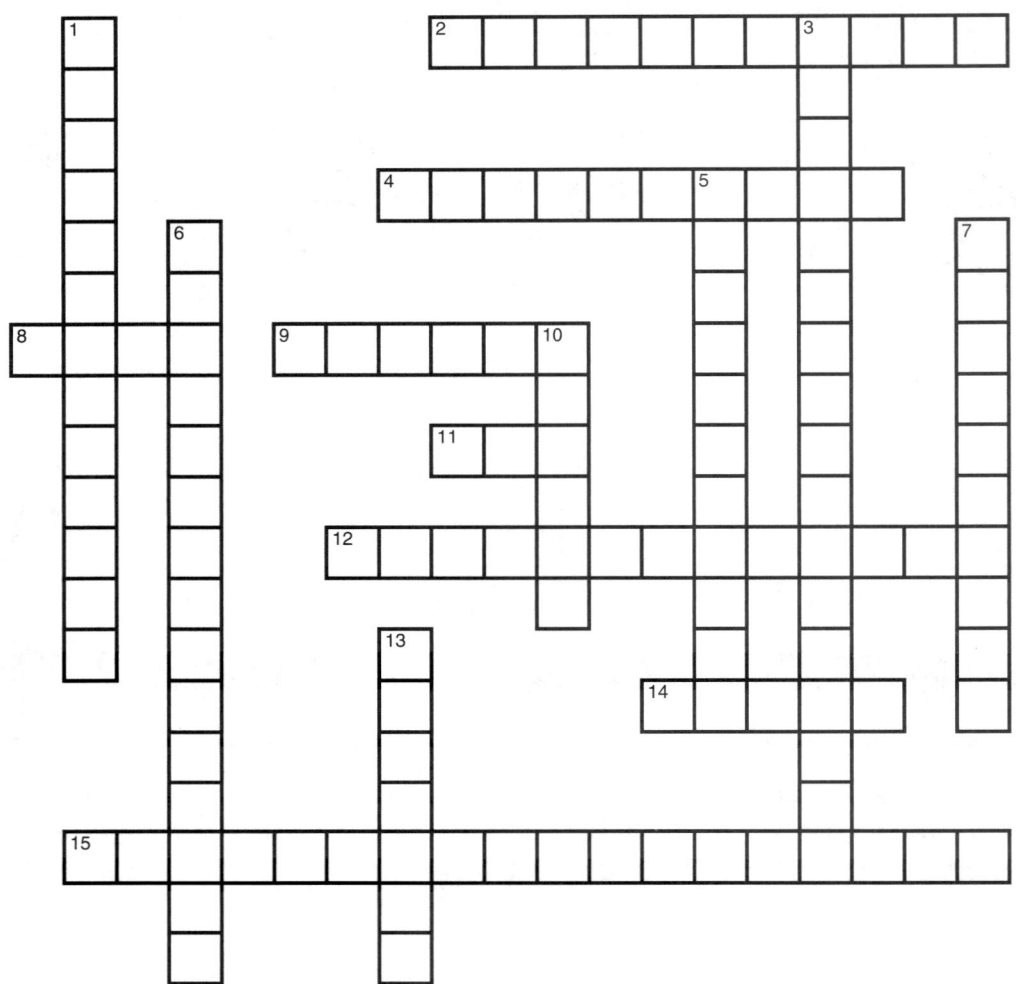

Across

2. Two conductors that are twisted together to reduce electrical interference.

4. A mechanism that turns the vehicle's turn signal and hazard flasher bulbs on and off.

8. A conductor usually made of multistranded copper with an external insulated coating.

9. The return path of electrical current in a vehicle chassis, other metal of the vehicle, or dedicated wire.

11. An abbreviation for amperes, the unit for current measurement.

12. A collection of wires or cables insulated from each other but bound together.

14. The unit for electrical frequency measurement.

15. The theory that electrons flow from positive to negative.

Down

1. Also called a pot, a three-terminal resistive device with one terminal connected to the input of the resistor, one terminal connected to the output of the resistor, and the third terminal connected to a moveable wiper arm that moves up and down the resistor.

3. A term used to describe a component that must be connected into a circuit with the current polarity to its terminals.

5. A type of flat harness in which cables are insulated from each other but joined together side by side.

6. A type of crimp or solder terminal that has an enclosed eyelet to connect the terminal with a bolt or screw.

7. A temperature-sensitive resistor. Its resistance varies substantially with temperature change.

10. The introduction of impurities to pure semiconductor materials to provide N- and P-type semiconductors.

13. A device used to measure current flow.

ASE-Type Questions

Read each item carefully, and then select the best response.

_____ 1. Tech A says that if the specified fuse keeps blowing, it is generally OK to replace it with a larger fuse. Tech B says that a fusible link is one type of circuit protection device. Who is correct?
   A. Tech A
   B. Tech B
   C. Both A and B
   D. Neither A nor B

_____ 2. Tech A says that the movement of electrons in a circuit is called current flow. Tech B says that the movement of electrons in a circuit is measured in amps. Who is correct?
   A. Tech A
   B. Tech B
   C. Both A and B
   D. Neither A nor B

_____ 3. Tech A says that electromotive force is also known as voltage. Tech B says that when electrons flow in one direction only, this is DC. Who is correct?
   A. Tech A
   B. Tech B
   C. Both A and B
   D. Neither A nor B

_____ 4. Tech A says that 18 AWG wire is larger than 12 AWG wire. Tech B says that the larger the diameter of the conductor, the more electrical resistance it has. Who is correct?
   A. Tech A
   B. Tech B
   C. Both A and B
   D. Neither A nor B

_____ 5. Two technicians are discussing electron flow. Tech A says that in "conventional theory" current is believed to flow from positive to negative. Tech B says that in "electron theory" current is believed to flow from negative to positive. Who is correct?
   A. Tech A
   B. Tech B
   C. Both A and B
   D. Neither A nor B

_____ 6. Tech A says that hertz is the number of cycles per second. Tech B says that hertz is the amount of current flow produced by an alternator. Who is correct?
   A. Tech A
   B. Tech B
   C. Both A and B
   D. Neither A nor B

_____ 7. Two technicians are discussing a series circuit with four resistors of various resistances. Tech A says that current flow will be different in each resistor. Tech B says that current flow will be the same in each resistor. Who is correct?
   A. Tech A
   B. Tech B
   C. Both A and B
   D. Neither A nor B

_____ **8.** Tech A says that a voltage drop is typically used to find excessive resistance in a circuit. Tech B says that high resistance creates heat at the point of resistance in the circuit. Who is correct?

    A. Tech A

    B. Tech B

    C. Both A and B

    D. Neither A nor B

_____ **9.** Tech A says that in a series circuit with two resistors of 120 ohms each, the total circuit resistance is 240 ohms. Tech B says that in a parallel circuit with two resistors of 120 ohms each, the total circuit resistance is 240 ohms. Who is correct?

    A. Tech A

    B. Tech B

    C. Both A and B

    D. Neither A nor B

_____ **10.** Tech A says that a relay is a one-way electrical check valve used in alternators to change AC into DC. Tech B says that a relay uses electromagnetism to open or close a switch. Who is correct?

    A. Tech A

    B. Tech B

    C. Both A and B

    D. Neither A nor B

Meter Usage and Circuit Diagnosis

Chapter Review

The following activities have been designed to help you refresh your knowledge of this chapter. Your instructor may require you to complete some or all of these activities as a regular part of your training program. You are encouraged to complete any activity that your instructor does not assign as a way to enhance your learning.

Matching

Match the following terms with the correct description or example.

A. Digital volt-ohmmeter (DVOM)

B. High resistance

C. Min/max setting

D. Open circuit

E. Probing technique

F. Short to power

_____ **1.** A setting on a DVOM to display the maximum and minimum readings.

_____ **2.** A term that describes a circuit or components with more resistance than designed.

_____ **3.** A condition in which current flows from one circuit into another.

_____ **4.** A circuit that has a break that prevents current from flowing.

_____ **5.** The way in which test probes are connected to a circuit.

_____ **6.** A test instrument with a digital display for measuring voltage, resistance, and current.

Multiple Choice

Read each item carefully, and then select the best response.

_____ **1.** A digital volt-ohmmeter is used to measure _____.

    A. voltage

    B. amperage

    C. resistance

    D. all of the above

_____ **2.** A digital volt-ohmmeter for three-phase fixed equipment and single-phase commercial lighting would have a _____ rating.

    A. CAT I

    B. CAT II

    C. CAT III

    D. CAT IV

_____ **3.** The _____ setting is often used to measure vehicle battery voltage while the engine is cranking or the battery is charging.

    A. hold

    B. sample

    C. min/max

    D. save

_____ **4.** The red lead for the DVOM is labeled with the _____ symbol.

    A. +

    B. V/Ω

    C. −

    D. ±

_____ **5.** All of the following are examples of common types of DVOM probes, _except:_
  A. ground probes
  B. alligator clips
  C. fine-pin probes
  D. insulation piercing probes

_____ **6.** The sum of all the _____ in a series circuit equals the supply voltage.
  A. resistors
  B. batteries
  C. voltage drops
  D. loads

_____ **7.** The _____ is the same in all parts of a properly working series circuit.
  A. voltage
  B. current
  C. resistance
  D. all of the above

_____ **8.** A(n) _____ must be connected in series within the circuit.
  A. DVOM
  B. resistor
  C. conductor
  D. ammeter

_____ **9.** As the position of the wiper of a potentiometer is changed, so is the _____ to a load connected to the potentiometer.
  A. voltage
  B. resistance
  C. current flow
  D. both A and C

_____ **10.** _____refers to a condition where power from one circuit leaks into another circuit.
  A. Short to power
  B. Grounds
  C. High resistance
  D. Shorts

True/False

If you believe the statement to be more true than false, write the letter "T" in the space provided. If you believe the statement to be more false than true, write the letter "F".

_____ **1.** Hybrid vehicles typically require meters and test leads rated as CAT III or CAT IV.

_____ **2.** 5208 MV is the same as 5208 millivolts. It could also be called 52.08 volts.

_____ **3.** The red lead is the positive lead, and the black lead is the negative lead.

_____ **4.** After back-probing always reinsulate the hole that the probe makes with room temperature vulcanizing silicone to prevent any corrosion.

_____ **5.** To accurately measure the resistance of a component, you should remove or isolate the component from the circuit.

_____ **6.** A DVOM reading of OL means overload and indicates the voltage being read is higher than the maximum allowed for the range.

_____ **7.** Voltage drop can be measured across components, connectors, or cables, but current has to be flowing to get an accurate measurement.

_____ **8.** Parallel circuits are commonly used in the vehicle's electrical system, especially for lights.

_____ **9.** A series-parallel circuit occurs when wanted resistance shows up in series with a parallel circuit.

_____ **10.** If the resistance and voltage of a circuit are known, then the theoretical current can be calculated using Ohm's law.

Fill in the Blank

Read each item carefully, and then complete the statement by filling in the missing word(s).

1. A basic digital _____-_____ can measure alternating current (AC) and direct current (DC) voltage, AC and DC amperage, and resistance.

2. _____ _____ are used to connect the DVOM to or into the circuit being tested and come in pairs: one red, the other black.

3. When the _____ function is activated, the display will hold the value on the display until the function or DVOM is turned off.

4. _____-_____ occurs when the probe is pushed in from the back of a connector to make a connection.

5. The _____ _____ fastens around the conductor and measures the strength of the magnetic field produced from current flowing through the conductor.

6. Unwanted voltage _____ in vehicle circuits can cause real problems and faults.

7. Current flow is _____ proportional to resistance.

8. The sum of the current flow in each branch is equal to the total _____ circuit current flow.

9. _____ _____ are used to extend connections to allow circuit readings or tests to be undertaken with a DVOM, an oscilloscope, current clamps on fuses, relays, and connector plugs on components.

10. _____ _____ can occur anywhere in the circuit and can be difficult to locate, especially if it is intermittent.

Labeling

Label the following diagrams with the correct terms.

1. Measuring voltage and current flow in a circuit with a single resistor and a 12-volt DC supply:

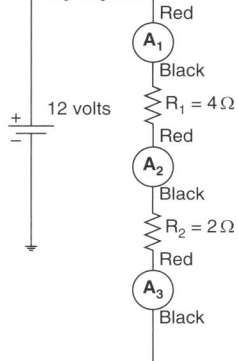

A. Current flow at A_1: _____.

B. Voltage at V_1: _____.

2. Measuring amperage in a circuit with two resistors in series with a 12-volt DC supply:

A. Current flow at A_1: _____.

B. Current flow at A_2: _____.

C. Current flow at A_3: _____.

3. Measuring amperage in a circuit with a relay controlled by a switch and a single resistor with a 12-volt DC supply:

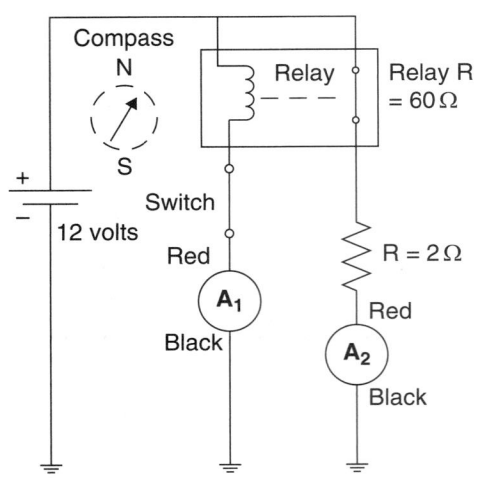

A. Current flow at A_1: _____.

B. Current flow at A_2: _____.

4. Measuring volts, amps, and ohms in a circuit with a 100 ohm resistor and a 12-volt DC supply:

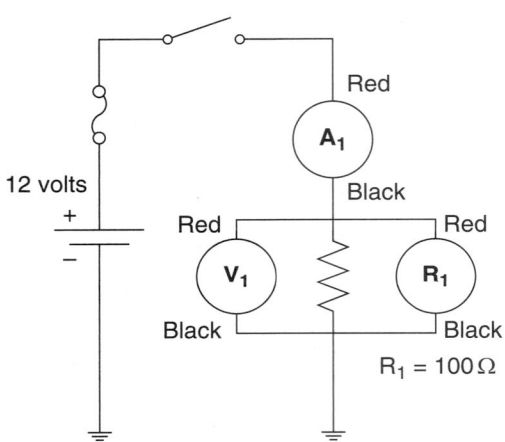

A. Current flow at A_1: _____.

B. Voltage at V_1: _____.

C. Resistance at R_1: _____.

5. Measuring volts, amps, and ohms in a circuit with a 100 ohm resistor and a 12-volt DC supply in series with an LED:

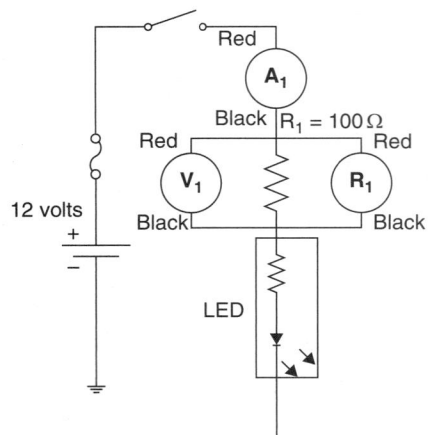

A. Current flow at A_1: _____.

B. Voltage at V_1: _____.

C. Resistance at R_1: _____.

6. Measuring amperage in a circuit with three resistors in parallel:

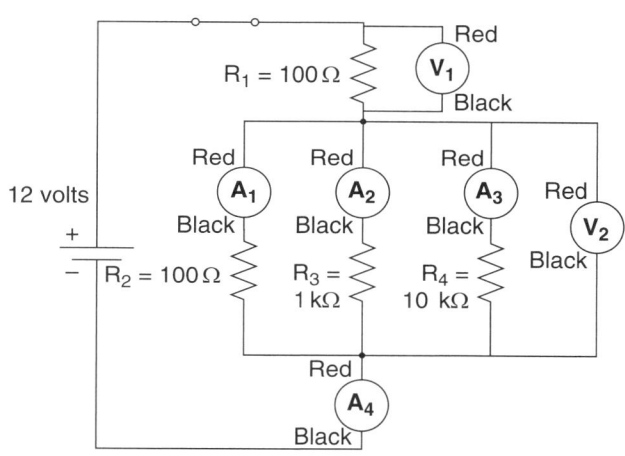

A. Current flow at A_1: _____.

B. Current flow at A_2: _____.

C. Current flow at A_3: _____.

D. Current flow at A_4: _____.

7. Series-parallel circuit formed by resistors R_1, R_2, R_3, and R_4:

A. Voltage at V_1: _____.

B. Voltage at V_2: _____.

C. Current flow at A_1: _____.

D. Current flow at A_2: _____.

E. Current flow at A_3: _____.

F. Current flow at A_4: _____.

Skill Drills

Test your knowledge of skill drills by filling in the correct words in the photo captions.

1. Checking Circuit Waveforms:

Step 1: Determine the circuit to be tested and the likely _____ and frequency of the _____ to be measured. Set the voltage level and _____ _____.

Step 2: Connect the _____ to the _____ in the circuit to be measured.

Step 3: _____ waveforms from the circuit being tested. _____ the waveform, comparing it to the manufacturer's specifications or _____ good waveforms.

2. Using a DVOM to Measure Voltage:

Step 1: Prepare the DVOM for testing _____ by connecting the _____ lead to the COM terminal and the _____ lead to the Volt/Ohms (V/_) terminal.

Step 2: Turn the _____ dial until you have selected the _____ for volts DC. The reading on the DVOM should now be at _____.

Step 3: Connect the black lead to the _____ battery post and the red lead to the _____ post. Measure the _____, and interpret the results.

3. Checking a Circuit with a Test Light:

Step 1: Connect the end of the light with the _____ on it to the negative battery _____. Touch the _____ end of the test light to the positive battery terminal. The _____ should come on.

Step 2: Connect the clip to a known good _____. A typical known good ground is any unpainted _____ surface on the vehicle that is directly _____ to the battery ground return system.

Step 3: Place the _____ on the terminal to be _____. If _____ is present, the light will come on.

Crossword Puzzle

Use the clues to complete the puzzle.

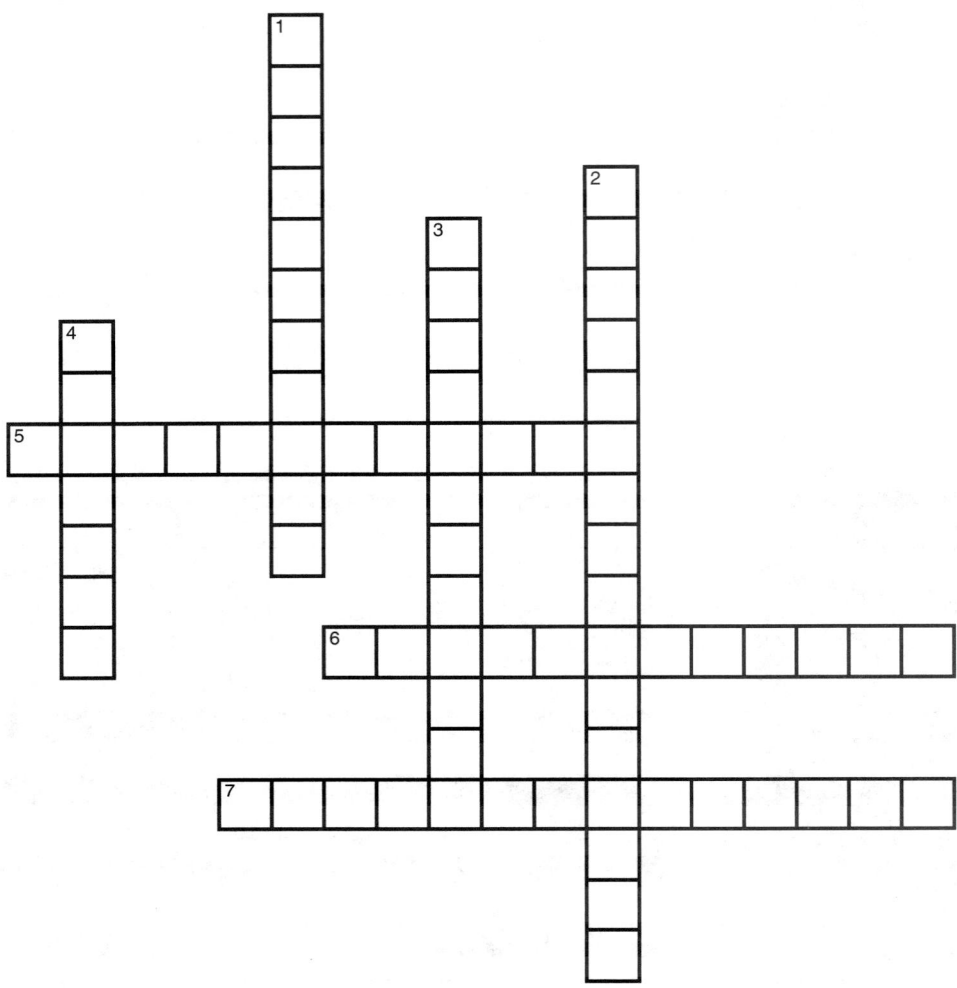

Across

5. A setting on a DVOM to store the present reading.

6. A condition in which the current flows along an unintended route.

7. A term that describes a circuit or components with more resistance than designed.

Down

1. A circuit that has a break that prevents current from flowing.

2. The way in which test probes are connected to a circuit.

3. A condition in which current flows from one circuit into another.

4. Fault conditions in a circuit where the circuit is unintentionally contacting a grounded component or wire. This may result in a short, in the case of power wire, or it could cause a circuit to stay line in the case of a switched ground circuit.

ASE-Type Questions

Read each item carefully, and then select the best response.

_____ **1.** Tech A says that total resistance goes up as more parallel paths are added. Tech B says that total amperage goes up as more parallel paths are added. Who is correct?
  A. Tech A
  B. Tech B
  C. Both A and B
  D. Neither A nor B

_____ **2.** Tech A says that when reading DC voltage on a meter, a "+" before the number means that there is a higher voltage at the red lead than the black lead. Tech B says that a "–" before the number means that there is a lower voltage at the red lead than the black lead. Who is correct?
  A. Tech A
  B. Tech B
  C. Both A and B
  D. Neither A nor B

_____ **3.** Tech A says that to read amperage the meter needs to be hooked up in series in a circuit. Tech B says that to read amperage at a load, place one lead of the ammeter on the input side of the load and the other lead on the output. Who is correct?
  A. Tech A
  B. Tech B
  C. Both A and B
  D. Neither A nor B

_____ **4.** Tech A says that when checking amperage on a battery, hook the red lead to positive and the black lead to ground. Tech B says that a resistance reading on a light bulb requires the DVOM to be hooked to each side of the bulb and the switch turned on. Who is correct?
  A. Tech A
  B. Tech B
  C. Both A and B
  D. Neither A nor B

_____ **5.** Tech A says that when checking a voltage drop across an open switch, a measurement of 12 volts means the circuit is open. Tech B says that when checking voltage drop across an open switch, a measurement of 12 volts means the switch contacts are closed. Who is correct?
  A. Tech A
  B. Tech B
  C. Both A and B
  D. Neither A nor B

_____ **6.** A customer complains of slow engine cranking. Tech A says that the starter is faulty and should be replaced. Tech B says that performing a voltage drop test on the high current side of the starter circuit is a valid test in this situation. Who is correct?
  A. Tech A
  B. Tech B
  C. Both A and B
  D. Neither A nor B

_____ **7.** Tech A says that when resistance increases, current flow decreases. Tech B says that when voltage decreases, current increases. Who is correct?
  A. Tech A
  B. Tech B
  C. Both A and B
  D. Neither A nor B

_____ **8.** Tech A says that an open circuit will typically cause higher current flow, which will blow the fuse. Tech B says that a short to ground can typically be found by checking voltage at different points in the circuit. Who is correct?
   A. Tech A
   B. Tech B
   C. Both A and B
   D. Neither A nor B

_____ **9.** Tech A says that when a fuse has popped, it just needs to be replaced, since it performed its job. Tech B says that when a fuse has blown, the circuit needs to be diagnosed because the fuse was probably not the problem. Who is correct?
   A. Tech A
   B. Tech B
   C. Both A and B
   D. Neither A nor B

_____ **10.** Technician A says that the black voltmeter lead stays on the negative battery terminal whenever you perform a "direct voltage drop test." Technician B says that an "indirect voltage drop test" is used to accurately measure current flow in a circuit. Who is correct?
   A. Tech A
   B. Tech B
   C. Both A and B
   D. Neither A nor B

Battery Systems

Chapter Review

The following activities have been designed to help you refresh your knowledge of this chapter. Your instructor may require you to complete some or all of these activities as a regular part of your training program. You are encouraged to complete any activity that your instructor does not assign as a way to enhance your learning.

Matching

Match the following terms with the correct description or example.

A. Absorbed glass mat
B. Battery terminal configuration
C. Cold cranking amps (CCA)
D. Current clamp

E. Cranking amps (CA)
F. Electrolyte
G. Keep alive memory (KAM)
H. Thermal runaway

_____ 1. The liquid in lead-acid battery cells. It is a mixture of about 67% water and 33% sulfuric acid.

_____ 2. A standard for rating the ability of a vehicle battery to supply high current under cold operating conditions.

_____ 3. The placement of positive and negative battery terminals.

_____ 4. A certain minimum amount of parasitic current draw that is used by the vehicle's circuits to maintain memory functions and monitor systems.

_____ 5. A device that clamps around a conductor to measure current flow. It is often used in conjunction with a DVOM.

_____ 6. A cycle during battery charging in which heating lowers resistance, which in turn increases current flow, which in turn further increases hear created. During this cycle, dangerous gases may build up, creating the potential for an explosion or damage through excessive current flow.

_____ 7. A type of lead-acid battery.

_____ 8. A standard similar to CCA but that measures the battery's function at a higher temperature—32°F (0°C).

Multiple Choice

Read each item carefully, and then select the best response.

_____ 1. A standard 12-volt car battery consists of _____ cells connected in series.
  A. 1
  B. 2
  C. 6
  D. 12

_____ 2. All of the following are methods used to rate automotive battery capacity, _except_:
  A. cold cranking amps
  B. voltage output
  C. cranking amps
  D. reserve capacity

_____ 3. A(n) _____ battery typically has no removable cell covers, so you cannot adjust or test the fluid levels inside.
  A. deep cycle
  B. absorbed glass mat
  C. low-maintenance
  D. reserve capacity

_____ **4.** Which of the following is a type of rechargeable cell battery?
  A. Lead-acid
  B. Nickel-cadmium
  C. Lithium ion
  D. All of the above

_____ **5.** All of the following are advantages of lithium-ion batteries, _except_:
  A. high energy density
  B. shelf life
  C. low internal resistance
  D. low self-discharge

_____ **6.** Which tool is commonly used to measure parasitic draw from the battery?
  A. Low-amp clamp
  B. Scan tool
  C. Ammeter
  D. Either A or C

_____ **7.** Performing a battery state-of-charge test with a(n) _____ will give a good indication of whether the battery needs to be charged or not.
  A. refractometer
  B. hydrometer
  C. DVOM
  D. any of the above

_____ **8.** The higher the _____, the higher the percentage of acid in the electrolyte, which corresponds to a high battery state of charge.
  A. specific gravity
  B. water level
  C. temperature
  D. amperage

True/False

If you believe the statement to be more true than false, write the letter "T" in the space provided. If you believe the statement to be more false than true, write the letter "F".

_____ **1.** Batteries store electricity in chemical form.

_____ **2.** The sulfuric acid contained in batteries is highly corrosive and can also be very harmful to metal, painted surfaces, and skin.

_____ **3.** Sizing has to do with the electrical specifications of the battery, while ratings have to do with the battery's physical attributes.

_____ **4.** Reserve capacity is the time in minutes that a new fully charged battery at 80°F (27°C) will supply a constant load of 25 amps without its voltage dropping below 10.5 volts for a 12-volt battery.

_____ **5.** The typical cell voltage of a lithium-ion battery is 1.2 volts.

_____ **6.** Installing a battery into a vehicle backward can instantly destroy on-board electronics.

_____ **7.** When jump starting a vehicle the slave battery is connected in series to the host (discharged) battery in order to provide additional capacity to crank and start the vehicle.

_____ **8.** Many municipalities require battery recycling and levy a "core charge" on every new automotive battery sold.

_____ **9.** State-of-charge testing tells us how much capacity the battery has left.

_____ **10.** In some cases, it may be possible to use a 9-volt memory minder or memory saver to maintain the vehicle's memory while the vehicle battery is disconnected.

Fill in the Blank

Read each item carefully, and then complete the statement by filling in the missing word(s).

  1. Each cell of a fully charged "12-volt" battery has a nominal _____ volts, for a total of _____ volts.

2. The more plate _____ _____ there is, the higher the electrical capacity of the battery.

3. Because the electrolyte in _____ _____ _____ batteries is a gel, which does not spill, this type of battery is especially handy for rough handling or tipping.

4. The charging process increases the amount of _____ in the electrolyte, making the electrolyte stronger.

5. The _____ the battery temperature, the higher the rate of charging.

6. Lithium-ion batteries may suffer from _____ _____ and cell rupture if overheated or overcharged.

7. Battery cables and terminals are designed to carry _____ _____ currents that are required during cranking of the automotive engine.

8. Battery terminals are usually _____ or _____ onto the battery cables to ensure strong, low-resistance connections.

9. Always remove the _____ or ground terminal first when disconnecting battery cables.

10. Some manufacturers say that their batteries should not be load tested, and instead should be _____ tested.

Labeling

Label the following diagrams with the correct terms.

1. Components of a simple battery:

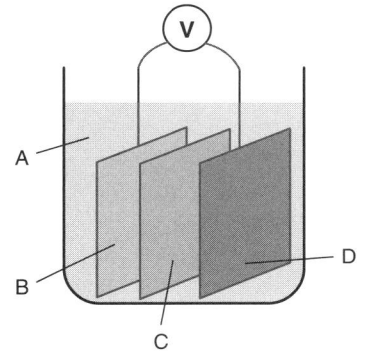

A. _____

B. _____

C. _____

D. _____

2. Typical plate arrangement in a wet cell battery:

A. _____

B. _____

C. _____

D. _____

E. _____

F. _____

G. _____

H. _____

I. _____

Crossword Puzzle

Use the clues to complete the puzzle.

Across

2. A certain minimum amount of parasitic current draw that is used by the vehicle's circuits to maintain memory functions and monitor systems.
6. Unwanted drain on the vehicle battery when the vehicle is off.
7. A standard used to specify the time in minutes that a battery will supply a load of 25 amps at 80°F (27°C) without its voltage dropping below 10.5 volts.
8. A device that clamps around a conductor to measure the battery's function at a higher temperature—32°F (0°C).
9. The escape of gas from the battery.
10. A temperature-sensitive resistor. Its resistance varies substantially with temperature change.

Down

1. The ability of a circuit or component to carry electrical loads.
3. A standard for rating the ability of a vehicle battery to supply high current under cold operating conditions.
4. The liquid in lead-acid battery cells. It is a mixture of about 67% water and 33% sulfuric acid.
5. A type of lead-acid battery.

ASE-Type Questions
Read each item carefully, and then select the best response.

_____ 1. Tech A says that a battery stores electrical energy in chemical form. Tech B says that a battery creates direct current. Who is correct?
  A. Tech A
  B. Tech B
  C. Both A and B
  D. Neither A nor B

_____ 2. Tech A says that a 12-volt battery has six cells. Tech B says that the more plates a cell in a battery has, the more voltage it creates. Who is correct?
  A. Tech A
  B. Tech B
  C. Both A and B
  D. Neither A nor B

_____ 3. Tech A says that a parasitic draw is measured in volts. Tech B says that pulling fuses one at a time can help locate a parasitic draw. Who is correct?
  A. Tech A
  B. Tech B
  C. Both A and B
  D. Neither A nor B

_____ 4. Tech A says that when disconnecting the battery, the negative terminal should be disconnected first. Tech B says that baking soda and water will remove oxidation from battery terminals. Who is correct?
  A. Tech A
  B. Tech B
  C. Both A and B
  D. Neither A nor B

_____ 5. Tech A says that checking the specific gravity will indicate the battery's cold cranking amps. Tech B says that a battery load test should be performed when the battery is heavily discharged. Who is correct?
  A. Tech A
  B. Tech B
  C. Both A and B
  D. Neither A nor B

_____ 6. Tech A says that before attempting to jump-start a vehicle, you need to make sure the battery is not frozen. Tech B says that it is good practice to place a load across the discharged battery, such as turning on the headlamps, to absorb any sudden rise in voltage. Who is correct?
  A. Tech A
  B. Tech B
  C. Both A and B
  D. Neither A nor B

_____ 7. Tech A says that there is an additional method to measure parasitic draw using an ohmmeter. Tech B says that the method is known as the Chesney parasitic load test after its creator, Sean Chesney. Who is correct?
  A. Tech A
  B. Tech B
  C. Both A and B
  D. Neither A nor B

_____ 8. Tech A says that a thermal runaway can occur during battery charging in which heating increases resistance and decreases the potential for an explosion or damage through excessive current flow. Tech B says that a thermal runaway can occur during battery charging in which heating lowers resistance and increases the potential for an explosion or damage through excessive current flow. Who is correct?
  A. Tech A
  B. Tech B
  C. Both A and B
  D. Neither A nor B

_____ **9.** Tech A says that some manufacturers recommend removing the negative battery terminal cable while charging a battery to reduce risk to the vehicle's electronics. Tech B says that this not necessary since all electronics are protected. Who is correct?

   A. Tech A
   B. Tech B
   C. Both A and B
   D. Neither A nor B

_____ **10.** Tech A says that cold cranking amps (CCA) ratings are a standard for determining the ability of a vehicle battery to supply high current under cold operating conditions. Tech B says that cranking amps (CA) is similar to CCA, but measures the battery's function at a higher temperature—32°F (0°C). Who is correct?

   A. Tech A
   B. Tech B
   C. Both A and B
   D. Neither A nor B

Starting and Charging Systems

Chapter Review

The following activities have been designed to help you refresh your knowledge of this chapter. Your instructor may require you to complete some or all of these activities as a regular part of your training program. You are encouraged to complete any activity that your instructor does not assign as a way to enhance your learning.

Matching

Match the following terms with the correct description or example.

A. Counter-electromotive force (CEMF)   **C.** Hold-in winding
B. Current clamp   **D.** Pull-in winding

_____ **1.** A low-current winding found in starter solenoids that holds the plunger in the activated position.

_____ **2.** A device that clamps around a conductor to measure current flow. It is often used in conjunction with a DVOM.

_____ **3.** Voltage created in the field windings as the motor rotates, which opposes battery voltage and limits motor speed.

_____ **4.** A high-current winding found in starter solenoids that pulls the solenoid plunger into the activated position.

Multiple Choice

Read each item carefully, and then select the best response.

_____ **1.** The starter motor is mounted on the transmission or cylinder block in a position to engage a _____ around the outside edge of the engine flywheel, flex plate, or torque converter.
   A. magnet
   B. commutator
   C. ring gear
   D. sleeve

_____ **2.** In the direct-drive system, the starter drive is mounted directly on one end of the _____.
   A. flywheel
   B. armature shaft
   C. transmission
   D. reduction gear

_____ **3.** The commutator end frame carries copper-impregnated carbon _____, which conduct current through the armature when it is being rotated in operation.
   A. spur gears
   B. brushes
   C. coils
   D. strips

_____ **4.** A(n) _____ consists of two semicircular segments that are connected to the two ends of the loop and are insulated from each other.
   A. helix
   B. armature
   C. commutator
   D. electromagnet

_____ 5. The _____ is typically a cylindrical device mounted on the starter motor that switches the high current flow required by the starter motor on and off and engages the starter drive with the ring gear.
   A. solenoid
   B. armature
   C. commutator
   D. pinion

_____ 6. For many years, manufacturers have placed switches in _____ with the starter solenoid windings, which prevents the starter from being activated.
   A. parallel
   B. series
   C. series/parallel
   D. either A or C

_____ 7. The _____ prevents the starter motor from being driven by the engine once the engine starts, which would spin the armature faster than it could handle.
   A. flywheel
   B. starter control circuit
   C. pole shoe
   D. overrunning clutch

_____ 8. The _____ converts mechanical energy into electrical energy by electromagnetic induction.
   A. DC generator
   B. inverter
   C. alternator
   D. rectifier

_____ 9. The voltage potential induced by an AC generator is called _____.
   A. electromotive force
   B. electromagnetic induction
   C. counter-electromotive force
   D. either A or B

_____ 10. To change AC to DC, automotive alternators use a rectifier assembly consisting of _____ in a specific configuration.
   A. transistors
   B. diodes
   C. resistors
   D. semiconductors

True/False

If you believe the statement to be more true than false, write the letter "T" in the space provided. If you believe the statement to be more false than true, write the letter "F".

_____ 1. All vehicles equipped with an automatic transmission use a neutral safety switch or a similar device.

_____ 2. The starter motor converts mechanical energy to electrical energy for the purpose of cranking the engine over.

_____ 3. A conductor loop that can freely rotate within the magnetic field is the most efficient motor design.

_____ 4. The hold-in winding draws a higher current and creates a stronger magnetic field than the pull-in winding.

_____ **5.** Some Ford vehicles use a separate starter relay in the engine compartment, instead of a solenoid, to control the high current for the starter motor.

_____ **6.** Some immobilizer systems now use keyless starting. The vehicle has a start button on the dash and does not require the key to be inserted into an ignition switch.

_____ **7.** A diode bridge gets its name from two diodes in series bridged with a wire.

_____ **8.** Rectification is a process of converting DC into the AC that is required by the battery and nearly all of the automobile systems.

_____ **9.** When the ignition is on, current flows from the positive battery terminal through the ignition switch and charge indicator lamp to the L terminal of the alternator.

_____ **10.** Alternators need to be replaced whenever they are electrically or mechanically faulty.

Fill in the Blank

Read each item carefully, and then complete the statement by filling in the missing word(s).

1. The _____ system provides a method of rotating the vehicle's internal combustion engine to begin the combustion cycle.

2. _____ _____ starters use an extra gear between the armature and the starter drive mechanism.

3. The _____ _____ _____ activates the solenoid winding to draw the plunger forward.

4. A(n) _____ _____ system is a computer-managed security system that disables the vehicle starter and engine systems by using an electronic system to uniquely identify each vehicle key by a security code system.

5. Most hybrid vehicles use a _____-_____ electric motor for engine start-up, auxiliary power, and regenerative braking functions.

6. A _____-_____ AC generator has only one stationary coil, which creates a single sine wave.

7. The _____ _____ monitors battery voltage and adjusts the current flow through the rotor appropriately.

8. Many alternators are designed with A-type circuits and a(n) _____ _____ on the regulator that provides a method of full-fielding the charging system with the engine running.

9. The _____ must not be operated with the battery disconnected or with the terminals at the back of the alternator disconnected.

10. _____ leads to electrical systems that do not function fully, and the battery may not fully charge, leading to an early death due to sulfation.

Labeling

Label the following diagrams with the correct terms.

1. Starter motor:

A. _____

B. _____

C. _____

D. _____

2. Series-wound starter:

A. _____

B. _____

C. _____

D. _____

E. _____

F. _____

G. _____

H. _____

I. _____

3. Simple multiloop motor and electromagnetic fields—with commutator and brushes:

A. _____

B. _____

C. _____

D. _____

E. _____

4. Hold-in and pull-in solenoid windings:

A. _____

B. _____

C. _____

D. _____

E. _____

F. _____

5. Ford moveable pole shoe starter:

A. _____

B. _____

C. _____

D. _____

E. _____

F. _____

G. _____

H. _____

I. _____

J. _____

K. _____

L. _____

M. _____

6. Basic starter control circuit:

Automatic Transmission

A. _____

B. _____

C. _____

D. _____

E. _____

F. _____

Manual Transmission

A. _____

B. _____

C. _____

D. _____

E. _____

7. Starter drive one-way clutch:

A. _____

B. _____

C. _____

D. _____

E. _____

F. _____

8. The alternator:

A. _____

B. _____

C. _____

D. _____

Crossword Puzzle

Use the clues to complete the puzzle.

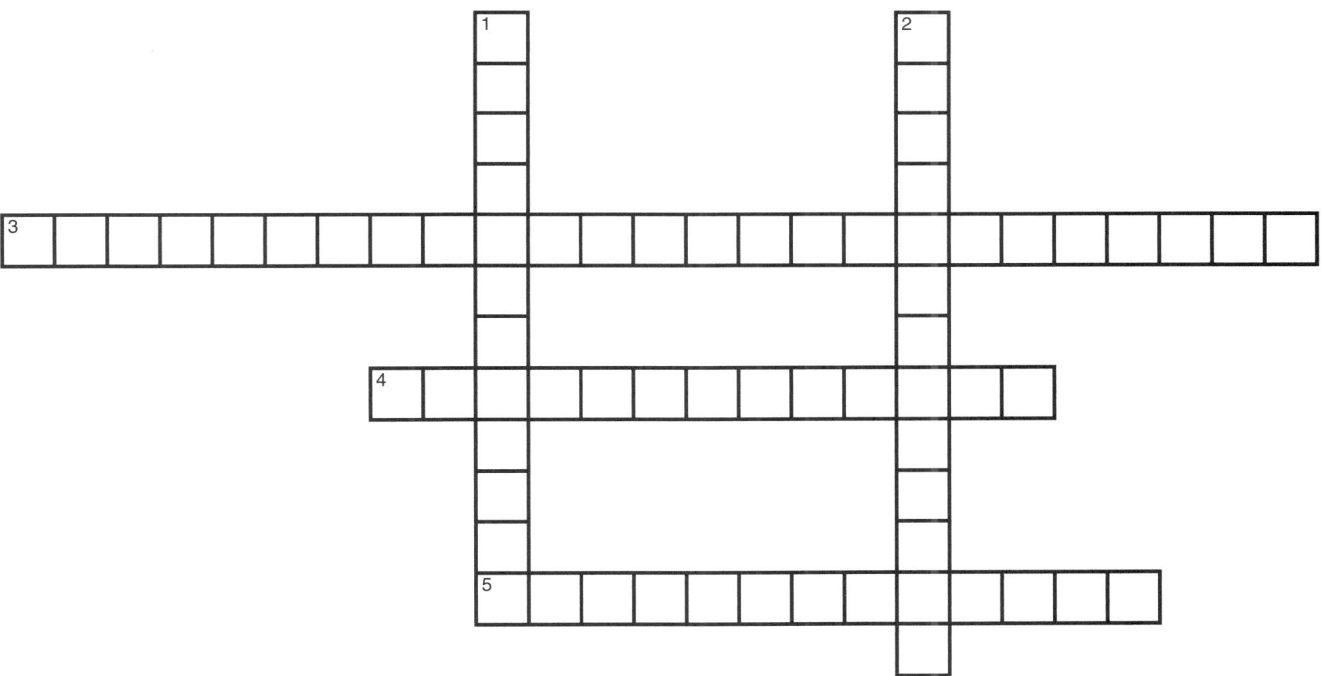

Across

3. Voltage created in the field windings as the motor rotates, which opposes battery voltage and limits motor speed.

4. A process of converting AC into DC required by the battery and nearly all of the automobile systems.

5. A high-current winding found in starter solenoids that pulls the solenoid plunger into the activated position.

Down

1. A device that clamps around a conductor to measure current flow. It is often used in conjunction with a DVOM.

2. A low-current winding found in starter solenoids that holds the plunger in the activated position.

ASE-Type Questions

Read each item carefully, and then select the best response.

_____ 1. Tech A says that some starters use gear reduction to improve efficiency. Tech B says that a starter converts electrical energy to mechanical energy. Who is correct?
  A. Tech A
  B. Tech B
  C. Both A and B
  D. Neither A nor B

_____ 2. Tech A says that the pull-in winding is short-circuited when the solenoid is fully engaged. Tech B says that the starter drive has a built in one-way clutch. Who is correct?
  A. Tech A
  B. Tech B
  C. Both A and B
  D. Neither A nor B

_____ 3. Tech A says that a voltage drop of 0.8 volts on the starter ground circuit is within specifications. Tech B says that high starter draw current could be caused by a spun main bearing in the engine. Who is correct?
  A. Tech A
  B. Tech B
  C. Both A and B
  D. Neither A nor B

_____ 4. Tech A says that the first item to check if an engine does not crank is the voltage to the S-terminal on the starter. Tech B says that the first item to check if an engine does not crank is battery voltage. Who is correct?
  A. Tech A
  B. Tech B
  C. Both A and B
  D. Neither A nor B

_____ 5. Tech A says that the voltage regulator controls the strength of the rotor's magnetic field. Tech B says that the voltage regulator is installed between the output terminal of the alternator and the positive terminal of the battery. Who is correct?
  A. Tech A
  B. Tech B
  C. Both A and B
  D. Neither A nor B

_____ 6. Tech A says that over charging can lead to short life of bulbs and other electrical devices. Tech B says that maintaining a proper charge is critical to long life and proper operation of the electrical system. Who is correct?
  A. Tech A
  B. Tech B
  C. Both A and B
  D. Neither A nor B

_____ 7. Tech A says that an excessive voltage drop in the alternator output circuit will likely cause a high charging system voltage. Tech B says that the alternator output circuit voltage drop must be checked with the charging system under heavy load. Who is correct?
  A. Tech A
  B. Tech B
  C. Both A and B
  D. Neither A nor B

_____ 8. Tech A says that most hybrid vehicles use the high voltage electric motor to start the engine. Tech B says that most hybrid engines need to crank over more slowly than regular ones. Who is correct?
  A. Tech A
  B. Tech B
  C. Both A and B
  D. Neither A nor B

_____ **9.** Tech A says that failure to crank over properly is always caused by mechanical problems. Tech B says that failure to crank over properly is always caused by electrical problems. Who is correct?
  A. Tech A
  B. Tech B
  C. Both A and B
  D. Neither A nor B

_____ **10.** Tech A says that the starter drive uses a pinion gear with an over running clutch. Tech B says that on some engines the starter motor is mounted on top of the engine under the intake manifold. Who is correct?
  A. Tech A
  B. Tech B
  C. Both A and B
  D. Neither A nor B

Lighting Systems

Chapter Review

The following activities have been designed to help you refresh your knowledge of this chapter. Your instructor may require you to complete some or all of these activities as a regular part of your training program. You are encouraged to complete any activity that your instructor does not assign as a way to enhance your learning.

Matching

Match the following terms with the correct description or example.

A. Halogen lamp
B. High-intensity discharge (HID)
C. Incandescent lamp
D. Light-emitting diode (LED)
E. Vacuum tube fluorescent (VTF)

_____ 1. A type of lighting that produces light with an electric arc rather than a glowing filament.

_____ 2. A type of lighting used for instrumentation displays on vehicle instrument panel clusters. This type of lighting emits a very bright light with high contrast and can shine in various colors. Also called vacuum fluorescent display (VFD).

_____ 3. A type of bulb that produces a bright white light.

_____ 4. A type of lighting used in various automotive applications, such as warning indicators and alphanumeric displays.

_____ 5. The traditional bulb that uses a heated filament to produce light.

Multiple Choice

Read each item carefully, and then select the best response.

_____ 1. Incandescent bulbs are inefficient, converting only about _____ of the electricity to visible light.
A. 10%
B. 15%
C. 25%
D. 30%

_____ 2. What kind of incandescent lamps are filled with bromine or iodine gas?
A. Incandescent
B. Vacuum tube fluorescent
C. Halogen
D. High-intensity discharge

_____ 3. LEDs are often required to give off a specified amount of light; to achieve this they are usually connected in groups called _____.
A. clusters
B. series strings
C. parallel sets
D. packs

_____ 4. Which type of bulb works well as a combination taillight and brake light because it has one filament that emits a small amount of light, and a second filament that emits more light?
A. LED cluster
B. Festoon-style bulb
C. Tandem bulb
D. Dual-filament bulb

_____ **5.** Which type of lamp base gets its name from the two retaining pins on the side of the base?
   A. Bayonet-style
   B. Festoon-style
   C. Wedge-style
   D. Pin-type

_____ **6.** Which type of lights usually light up as part of a self-test when the ignition initially comes on to show they are in working order?
   A. Marker lights
   B. Warning lamps
   C. Tail lights
   D. Courtesy lights

_____ **7.** For safety reasons, _____ continue to operate when the light switch is moved to the headlight position.
   A. park lights
   B. license plate illumination lamps
   C. tail lights
   D. both A and C

_____ **8.** Many vehicles by law now have a higher additional third brake light mounted on top of the trunk lid or on the rear window called a _____ light.
   A. pedestal light
   B. courtesy light
   C. center high mount stop light
   D. festoon light

_____ **9.** White lights mounted at the rear of a vehicle that provides the driver with vision behind the vehicle at night are called _____.
   A. back-up lights
   B. tail lights
   C. reverse lights
   D. either A or C

_____ **10.** Which type of lights simultaneously cause a pulsing in all exterior indicator lights and both indicator lights on the instrument panel?
   A. Hazard warning lights
   B. Turn signal lights
   C. Backup lights
   D. Driving lights

_____ **11.** Existing lights that turn on when the vehicle is running and turn off when the engine stops in order to improve the vehicle's visibility to other drivers in all weather conditions are called _____.
   A. hazard warning lights
   B. dual-filament lights
   C. courtesy lights
   D. daytime running lights

_____ **12.** A(n) _____ headlight has a highly polished aluminized glass reflector that is fused to the optically designed lens.
   A. dual-filament
   B. semi-sealed beam
   C. sealed-beam
   D. HID

_____ **13.** Which type of headlight is filled with xenon gas and does not use a filament-style lamp?
   A. LED
   B. HID
   C. Sealed-beam
   D. Halogen

_____ **14.** Which type of lights are used with other vehicle lighting in poor weather such as thick fog, driving rain, or blowing snow?
   A. Driving lights
   B. Cornering lights
   C. Fog lights
   D. Smart lights

_____ **15.** To improve visibility during night driving, some vehicle manufacturers provide _____.
   A. night vision
   B. driving lights
   C. fog lights
   D. cornering lights

True/False

If you believe the statement to be more true than false, write the letter "T" in the space provided. If you believe the statement to be more false than true, write the letter "F".

_____ **1.** Wattage is found by multiplying the voltage used by the lamp by the current flowing through it.

_____ **2.** High-intensity halogen lamp light comes from metallic salts that are vaporized within an arc chamber.

_____ **3.** Some countries mandate that HID headlamps may only be installed on vehicles with lens-cleaning systems and automatic self-leveling systems.

_____ **4.** Wattage is an indication of light output.

_____ **5.** Dome lights use festoon-style lights, which have a base on each end of a cylindrical light bulb.

_____ **6.** Government regulations control the height of taillights and their brightness.

_____ **7.** Taillights are the only white lights on the rear of the vehicle.

_____ **8.** Imported vehicles tend to have separate amber-colored turn signal lamps on both the front and the rear of the vehicle.

_____ **9.** The beam selector switch is a double-pole, single-throw switch, meaning it has two movable poles but makes contact in one position.

_____ **10.** Some vehicles use LED lights as headlights.

_____ **11.** Passive night vision systems use a heat-sensing camera to pick up thermal radiation emitted by objects.

_____ **12.** Cornering lights are white lights usually installed into the bumper or fender and are designed to provide side lighting when the vehicle is turning corners.

_____ **13.** Each lighting circuit has particular operating characteristics based on its purpose and the system design.

_____ **14.** HID lamps produce a very bright red light.

_____ **15.** Not all wiring diagrams use the same symbols or the same numbering system.

Fill in the Blank

Read each item carefully, and then complete the statement by filling in the missing word(s).

1. Modern vehicles use many different kinds and sizes of _____, also known in some places as light bulbs or light globes.

2. One of the advantages of a(n) _____ is that it turns on instantly.

3. In a bulb marked 12V/21W, the _____ will consume 21 watts of power when 12 volts is applied across it.

4. Many newer bulbs use a _____ base either made from the glass bulb itself or with a built-in plastic base.

5. _____ lights are usually controlled by the vehicle body computer with inputs from the ignition and door switches either in the handle, latch, or door pillar.

6. _____ lights are used to mark the sides of some vehicles and are often located down the sides of the vehicle or trailer.

7. Today, computer-controlled brake lights are activated by the _____ _____ _____ when the computer senses an input from the brake pedal switch.

8. The cancelling mechanism of the _____ _____ lights operates to return the switch to its central or "off" position after a turn has been completed and the steering wheel is returned to the straight-ahead position.

9. Some domestic vehicles use the rear brake lamps as turn signals by flashing the _____ _____ on one side to indicate the turn.

10. Most vehicle _____ require two beams to provide for a high beam and low beam operation.

11. A(n) _____-_____ headlight system produces a high-intensity forward beam and uses a lens system to project the light forward, rather than the traditional reflector system.

12. Active _____ _____ systems use an infrared light generator that projects infrared light in front of and to the side of the roadway ahead.

13. _____ _____ are installed on the front of the vehicle and provide higher intensity illumination over longer distances than standard headlight systems.

14. _____ _____ turn on only when the headlights and turn signal switches on, and turn off automatically when the turn signal cancels.

15. Manufacturers design the circuits to operate in specific ways and then create schematics or _____ _____, which are a diagrammatic layout of the entire circuit.

Labeling

Label the following diagrams with the correct terms.

1. Projector bulb assembly:

A. _____

B. _____

C. _____

D. _____

E. _____

F. _____

G. _____

2. Battery and fuse symbols:

A. _____

B. _____

C. _____

D. _____

3. Ground and connector symbols:

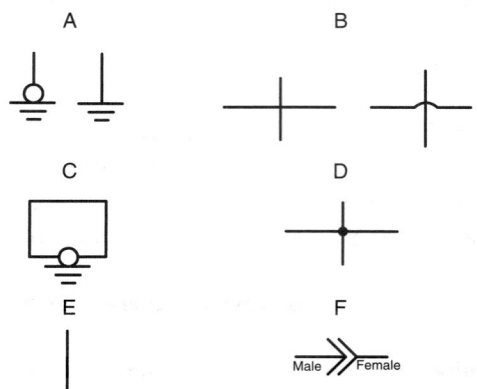

A. _____

B. _____

C. _____

D. _____

E. _____

F. _____

4. Switch symbols:

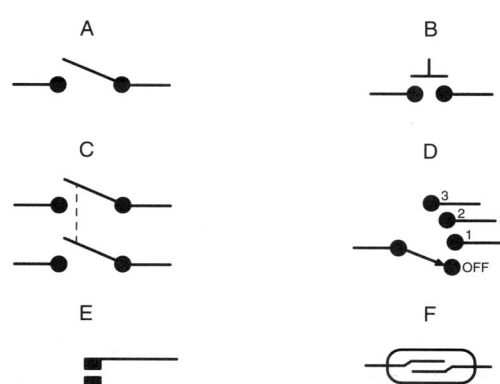

A. _____

B. _____

C. _____

D. _____

E. _____

F. _____

5. Resistors, coils, and relay symbols:

A. _____

B. _____

C. _____

D. _____

E. _____

F. _____

G. _____

H. _____

6. Semiconductor symbols:

A   B   C

D   E

F   G

A. _____

B. _____

C. _____

D. _____

E. _____

F. _____

G. _____

7. Capacitor and device symbols:

A   B

C   D

E   F

A. _____

B. _____

C. _____

D. _____

E. _____

F. _____

8. Motors, generators, and solenoid symbols:

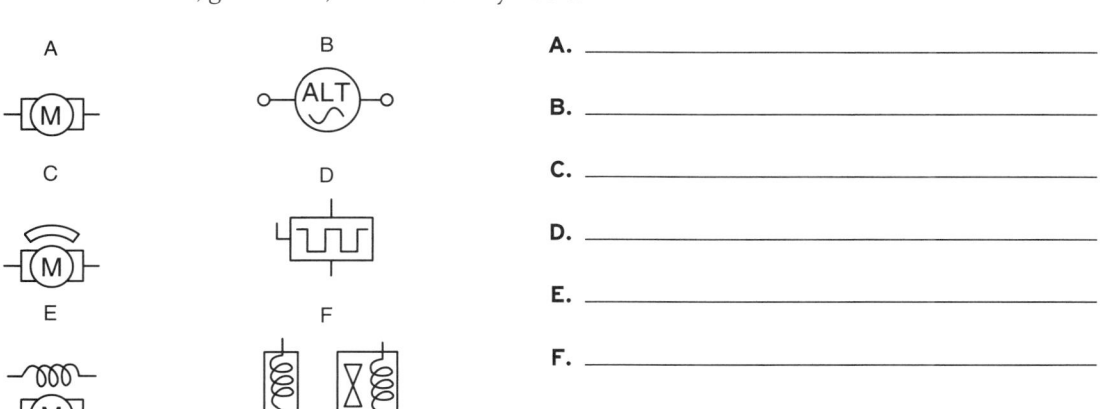

A   B

C   D

E   F

A. _____

B. _____

C. _____

D. _____

E. _____

F. _____

9. Gauges and warning device symbols:

A

B

C

D

E

F

A. _____

B. _____

C. _____

D. _____

E. _____

F. _____

Skill Drills

Test your knowledge of skill drills by filling in the correct words in the photo captions.

1. Checking and Changing an Exterior Light Bulb:

Step 1: Remove the _____ to expose the bulb. If the bulb is _____ mounted, gently grip the bulb and push it inward. Turn the bulb slightly _____, and remove it from the bulb holder.

Step 2: Inspect the bulb holder to make sure there is no _____. If there is, clean it with a bulb socket _____ _____ or _____ _____.

Step 3: Insert the new bulb into the bulb holder, _____ it fully, turn it slightly _____, and release it. Test it by _____ it on and off. Then replace the cover, and _____ it again.

2. Checking and Changing a Headlight Bulb:

Step 1: Test the vehicle headlights. Obtain the replacement _____ for the vehicle. Unplug the _____ _____ at the _____ of the lamp unit.

Step 2: _____ the old bulb, and _____ it with the new one. Handle the new bulb only by its _____ or, if supplied, by the _____ cover.

Step 3: Replace the unit and the _____ _____ or bulb assembly, and then plug in the _____. Switch on the lights again to _____ that they are both operating correctly.

3. Aiming Headlights:

Step 1: Make sure the tires are _____ properly, the _____ point straight ahead, and there is no extra _____ in the vehicle. Position the vehicle correctly in relation to the headlamp _____ unit following the equipment manufacturer's instructions. Calibrate the aligner for any _____ _____ and for the vehicle being tested.

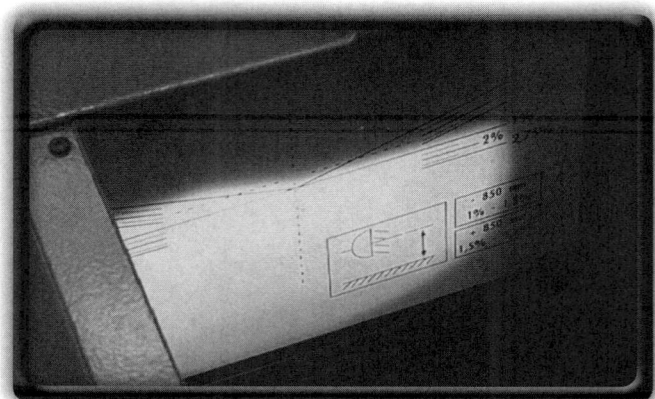

Step 2: On the types of _____ that require the headlights to be on during alignment, turn the headlights on to a _____ _____ setting. The _____ of the illuminating beams should be in the _____ _____ quadrants of the chart or wall markings or as specified by the manufacturer.

Step 3: The high beam should be _____, falling on the _____ of the horizontal and vertical marks or as specified by the manufacturer. If necessary, turn the _____ _____ on the headlight so the lights _____ to the correct places or _____ on the levels are centered, depending on the type of aligner equipment you are using.

Crossword Puzzle

Use the clues to complete the puzzle.

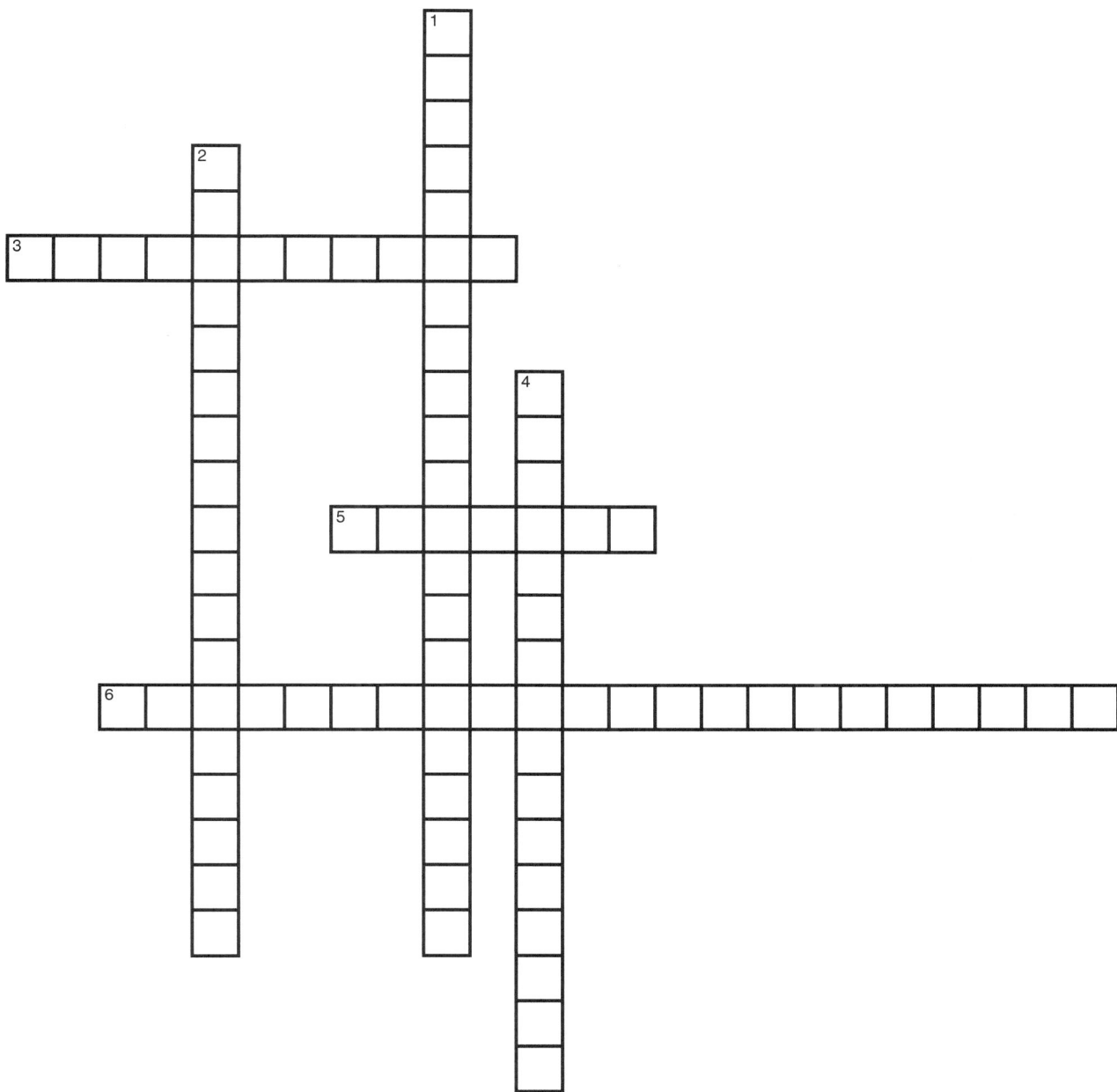

Across

3. A type of bulb that produces a bright white light.

5. A device that increases lighting voltage substantially and controls the current to the bulb.

6. A type of lighting that produces light with an electric arc rather than a glowing filament.

Down

1. A type of lighting used for instrumentation displays on vehicle instrument panel clusters. This type of lighting emits a very bright light with high contrast and can display in various colors.

2. A type of lighting various automotive applications, such as warning indicators and alphanumeric displays.

4. The traditional bulb that uses a heated filament to produce light.

ASE-Type Questions

Read each item carefully, and then select the best response.

_____ **1.** Tech A says that a voltage drop on the ground side of a bulb won't affect its brightness because the electricity has already been used up. Tech B says that a voltage drop is one way to determine if there is unwanted resistance in a circuit. Who is correct?
  A. Tech A
  B. Tech B
  C. Both A and B
  D. Neither A nor B

_____ **2.** Tech A says that incandescent bulbs resist vibration well. Tech B says that HID headlamps require up to approximately 25,000 volts to start. Who is correct?
  A. Tech A
  B. Tech B
  C. Both A and B
  D. Neither A nor B

_____ **3.** Tech A says that many automotive light bulbs have more than one filament inside. Tech B says that some lights use a filament made of quartz. Who is correct?
  A. Tech A
  B. Tech B
  C. Both A and B
  D. Neither A nor B

_____ **4.** Tech A says that LED brake lights illuminate faster than incandescent bulbs. Tech B says that LED brake lights have more visibility and last longer. Who is correct?
  A. Tech A
  B. Tech B
  C. Both A and B
  D. Neither A nor B

_____ **5.** Tech A says that if the brake light switch is open, neither brake light will illuminate. Tech B says that the back-up lights are connected in parallel with the tail lights. Who is correct?
  A. Tech A
  B. Tech B
  C. Both A and B
  D. Neither A nor B

_____ **6.** Tech A says that some brake lights get power from the brake switch through the turn signal switch. Tech B says many turn signals use amber lights. Who is correct?
  A. Tech A
  B. Tech B
  C. Both A and B
  D. Neither A nor B

_____ **7.** Tech A says that daytime running light systems illuminate the tail lights to enhance visibility. Tech B says that the CHMSL is illuminated when the tail lights are activated. Who is correct?
  A. Tech A
  B. Tech B
  C. Both A and B
  D. Neither A nor B

_____ **8.** Tech A says that some turn signals are flashed by a flasher can. Tech B says that some turn signals are flashed by the BCM. Who is correct?
  A. Tech A
  B. Tech B
  C. Both A and B
  D. Neither A nor B

_____ **9.** Tech A says that setting tire pressure is part of a headlight adjustment. Tech B says that some headlight aligners require the headlight to be on during the alignment process. Who is correct?

    A. Tech A

    B. Tech B

    C. Both A and B

    D. Neither A nor B

_____ **10.** Tech A says that if a bulb is dim when operated, the circuit likely has a short to ground. Tech B says that if a bulb is dim, you should perform a voltage drop test on the power and ground side of the bulb. Who is correct?

    A. Tech A

    B. Tech B

    C. Both A and B

    D. Neither A nor B

Body Electrical System

Tire Tread:
© AbleStock

Chapter Review

The following activities have been designed to help you refresh your knowledge of this chapter. Your instructor may require you to complete some or all of these activities as a regular part of your training program. You are encouraged to complete any activity that your instructor does not assign as a way to enhance your learning.

Matching

Match the following terms with the correct description or example.

A. Blower motor
B. Bus
C. Brushless DC motor
D. Controller area network (CAN)

E. Data link connector (DLC)
F. Permanent magnet electric motor
G. Stepper motor
H. Supplemental restraint system (SRS)

_____ 1. A type of brushless motor with a key difference: It is designed to rotate in fixed steps through a set number of degrees.

_____ 2. The port to which a scan tool can be connected.

_____ 3. An electric motor that does not have any brushes and is sometimes called an "electronically commutated motor." In this type of motor, an electronic control module replaces the brushes and commutator.

_____ 4. The data transport system for electronic control modules.

_____ 5. The network that connects the vehicle's electronic control modules.

_____ 6. A passenger safety system, such as airbags and seat belt pretensioners.

_____ 7. An electric motor, usually the permanent magnet type that moves air over the air-conditioning evaporator and heater core.

_____ 8. An electric motor in which the magnetic field in the casing is produced by permanent magnets, while the armature has an electromagnetic field generated by passing electrical current through loops or windings, thereby producing the motor action.

Multiple Choice

Read each item carefully, and then select the best response.

_____ 1. Which of the following is a high-speed network used for critical data such as supplemental restraint systems, antilock brake systems, and engine controls?
A. CAN-bus A
B. CAN-bus B
C. CAN-bus C
D. CAN-bus D

_____ 2. A group of eight bits of binary data is called a _____.
A. microbyte
B. byte
C. kilobyte
D. megabyte

_____ **3.** Which of the following is a standard CAN-bus network configuration?
   A. Bus parallel
   B. Star parallel
   C. Looped series
   D. All of the above

_____ **4.** What type of motor can be used to rotate a short distance and then stop or go in the reverse direction for a set number of degrees by controlling each coil individually through the microcontroller?
   A. Brush-type motor
   B. Potentiometer
   C. Stepper motors
   D. Induction motor

_____ **5.** The horn switch on vehicles with a driver's side airbag is usually mounted in the steering wheel and requires a _____ to maintain an electrical connection to the circuit as the steering wheel rotates.
   A. clock spring
   B. slot switch
   C. slip ring and brush assembly
   D. transmitter

_____ **6.** In keyless systems, _____ and key fobs transmit a coded signal that is received by the vehicle's thief-deterrent module.
   A. remote fob
   B. remote control keys
   C. teen driver key
   D. Either A or C

_____ **7.** What controls the intermittent operation of a windshield wiper timer circuit?
   A. Discrete electronic timer
   B. Time-delay cube relay
   C. Vehicle body computer
   D. Either A or C

_____ **8.** Which of the following is an example of a secondary vehicle safety system?
   A. Seat belt
   B. Collapsible steering column
   C. SRS airbag
   D. Seat belt pretensioners

_____ **9.** To prevent incorrect and unnecessary airbag deployment, systems include a(n) _____ mounted within the SRS control unit.
   A. mercury switch
   B. accelerometer
   C. safing sensor
   D. proximity sensor

_____ **10.** Which of the following can remote start systems control?
   A. Heating systems
   B. Air conditioning systems
   C. Window defrost systems
   D. All of the above

True/False

If you believe the statement to be more true than false, write the letter "T" in the space provided. If you believe the statement to be more false than true, write the letter "F".

_____ 1. CAN-bus-compliant diagnostic systems have been required on all vehicles sold in the United States since 2008.

_____ 2. Some networks transmit and receive signals over a single wire, but most have dual wires and are commonly called "CAN-bus high (H)" and "CAN-bus low (L)," with the same message sent on both lines.

_____ 3. A brushless DC motor has brushes and is sometimes called an "electronically commutated motor."

_____ 4. A stepper motor is designed to rotate in fixed steps through a set of number of degrees.

_____ 5. Blower motor speed can be controlled by using a number of resistors connected in series and a switch to select between combinations of resistors or a more complex electronic speed control module.

_____ 6. Cooling fans in modern vehicles are usually switched by relays, which are controlled by an electric control unit based on information from the coolant temperature sensor.

_____ 7. The clock spring is a device within the steering wheel with a flexible ribbon cable that can rotate endlessly in a single direction.

_____ 8. Power door locks use an electrical actuator to move the mechanical door lock mechanism.

_____ 9. The wiper motor will usually have a high and low speed and a time delay or intermittent operation.

_____ 10. There are two types of airbag triggering mechanisms: electrical and automatic.

Fill in the Blank

Read each item carefully, and then complete the statement by filling in the missing word(s).

1. In electrical and electronic systems, a _____ is a means of connecting many electrical or electronic components for either data or power sharing.

2. Regardless of how the network is physically connected, the CAN-bus-H and CAN-bus-L systems will have terminating resistors connected to the data lines to form a(n) _____ through the resistors.

3. The _____ _____ in a vehicle is usually the permanent magnet type and moves air over the air-conditioning evaporator and heater core.

4. _____ _____ in modern vehicles are usually switched by relays, which are controlled by an electronic control unit based on information from the coolant temperature sensor.

5. Legislation requires vehicles to have a working _____ and in most cases also prescribes the sound level required to be made by the horn.

6. The sound of a horn is produced by the _____ of a metal diaphragm, which is operated by an electromagnet switched by a set of contacts.

7. The driver's side door is usually a _____ door lock, which means that when it is locked or unlocked with the key (or the remote fob) the other locks follow suit.

8. Removing _____ _____ _____ requires the removal of the arm rest, door lever, window crank, and any switch panels mounted on the door.

9. Seat belt _____ are used to tighten the seat belt in a severe frontal accident.

10. _____ occurs when the theft-deterrent system prevents the vehicle's engine from starting or the transmission from operating without the properly authorized key.

Labeling
Label the following diagrams with the correct terms.

1. CAN-bus configurations:

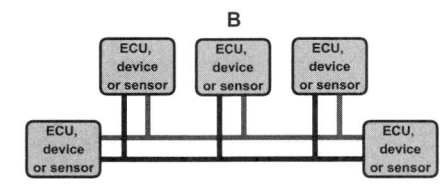

A. _____

B. _____

C. _____

2. Brush-type motor:

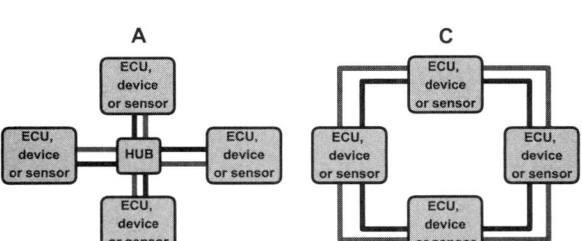

A. _____

B. _____

C. _____

D. _____

E. _____

3. Stepper motor:

A. _____

B. _____

C. _____

4. Parts of an airbag assembly:

A. _____

B. _____

C. _____

D. _____

5. Seat belt pretensioner:

A. _____

B. _____

C. _____

Crossword Puzzle

Use the clues to complete the puzzle.

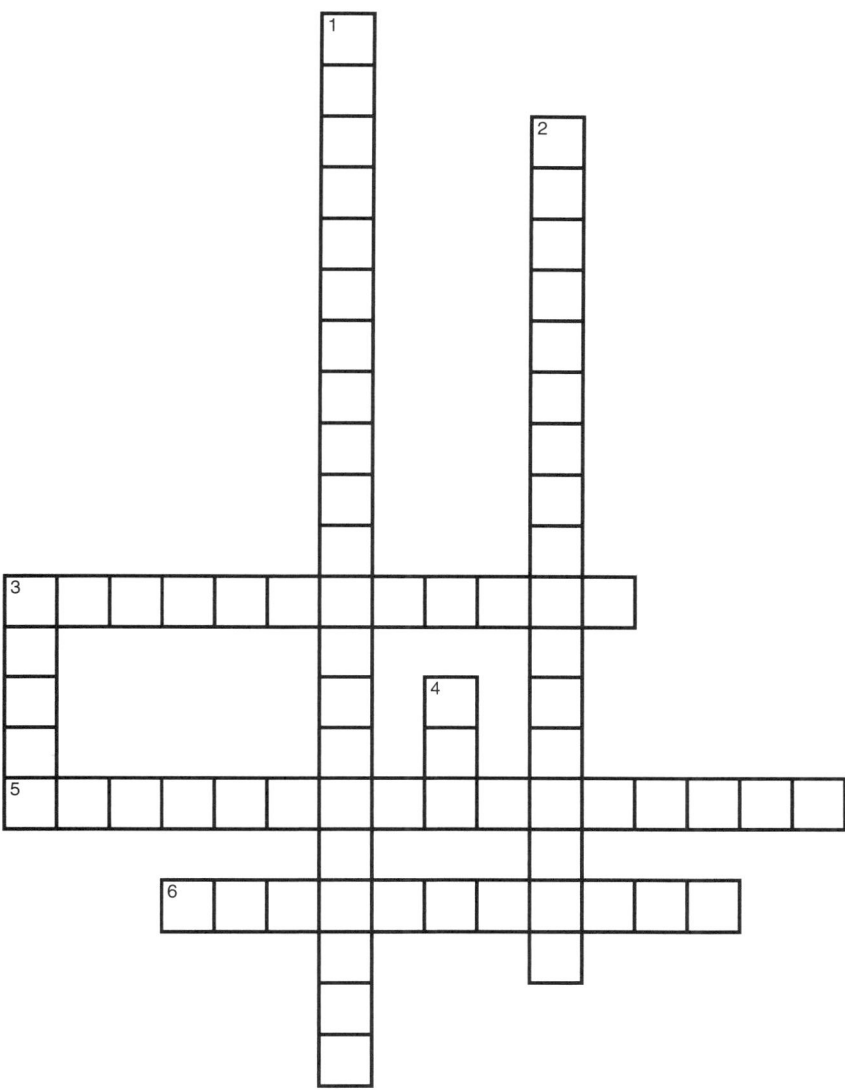

Across

3. A type of brushless motor with a key difference: It is designed to rotate in fixed steps through a set number of degrees.

5. An electric motor that does not have any brushes and is sometimes called an "electronically commutated motor." In this type of motor, an electronic control module replaces the brushes and commutator.

6. An electric motor, usually the permanent magnet type, that moves air over the air-conditioning evaporator and heater core.

Down

1. The network that connects the vehicle's electronic control modules.

2. The port to which a scan tool can be connected.

3. The component inside the airbag inflator that triggers the airbag deployment.

4. The date transport system for electronic control modules.

ASE-Type Questions

Read each item carefully, and then select the best response.

_____ **1.** Tech A says that CAN A is a low-speed data network. Tech B says that CAN C is a high-speed network. Who is correct?
- **A.** Tech A
- **B.** Tech B
- **C.** Both A and B
- **D.** Neither A nor B

_____ **2.** Tech A says that CAN C uses two 120-ohm resistors with a total circuit resistance of 60 ohms. Tech B says that CAN C uses two 120-ohm resistors with a total circuit resistance of 240 ohms. Who is correct?
- **A.** Tech A
- **B.** Tech B
- **C.** Both A and B
- **D.** Neither A nor B

_____ **3.** Tech A says that a twisted pair of wires in a wire loom makes it easier to route wires to their destination. Tech B says that the twisted pair of wires transmits signals of opposite polarity on the two lines. Who is correct?
- **A.** Tech A
- **B.** Tech B
- **C.** Both A and B
- **D.** Neither A nor B

_____ **4.** Two technicians are discussing windshield washers that aren't working. Tech A says that the windshield washer motor could be faulty. Tech B says that the washer nozzles could be plugged up. Who is correct?
- **A.** Tech A
- **B.** Tech B
- **C.** Both A and B
- **D.** Neither A nor B

_____ **5.** Tech A says that stepper motors are used as blower motors to achieve various speeds controlled by the driver. Tech B says that stepper motors are used where precise movement must occur, such as in electronic throttle controls. Who is correct?
- **A.** Tech A
- **B.** Tech B
- **C.** Both A and B
- **D.** Neither A nor B

_____ **6.** Tech A says that some airbags are dual stage and can have two charges in them. Tech B says that airbags must be manually deflated after being deployed. Who is correct?
- **A.** Tech A
- **B.** Tech B
- **C.** Both A and B
- **D.** Neither A nor B

_____ **7.** Tech A says that the clock spring is a device used in the steering wheel to transmit signals across the rotating electrical connection. Tech B says that the clock spring must be wound correctly during installation, as it only allows turning both ways a certain number of rotations. Who is correct?
- **A.** Tech A
- **B.** Tech B
- **C.** Both A and B
- **D.** Neither A nor B

_____ **8.** Tech A says that most SRS systems use a safing sensor to reduce the possibility of accidental deployment of the airbags. Tech B says that airbags are designed to act like a nice soft pillow in an accident. Who is correct?
- **A.** Tech A
- **B.** Tech B
- **C.** Both A and B
- **D.** Neither A nor B

_____ **9.** Tech A says that if airbags have been deployed, the affected seat belts and pre-tensioners need to be replaced. Tech B says that there are relatively large holes in the rear of an airbag that allow it to quickly deflate after deployment. Who is correct?

A. Tech A

B. Tech B

C. Both A and B

D. Neither A nor B

_____ **10.** Tech A says that in some vehicles, airbag deployment can be based on the speed of the impact and the weight of the driver. Tech B says that the SRS seat belt pre-tensioners tighten the seat belt when you connect the seat belt. Who is correct?

A. Tech A

B. Tech B

C. Both A and B

D. Neither A nor B

Principles of Heating and Air-Conditioning Systems

Chapter Review

The following activities have been designed to help you refresh your knowledge of this chapter. Your instructor may require you to complete some or all of these activities as a regular part of your *Fundamentals of Automotive Technology* training program. You are encouraged to complete any activity that your instructor does not assign as a way to enhance your learning.

Matching

Match the following terms with the correct description or example.

A. Accumulator
B. Atmospheric pressure
C. British thermal unit (Btu)
D. Chlorofluorocarbon (CFC)
E. Closed-loop system
F. Condensation
G. Convection

H. Dichlorodifluoromethane (R-12)
I. Electric servo
J. Evaporator
K. Fin
L. Fixed orifice tube system
M. Heat transfer
N. Latent heat of condensation

O. Latent heat of evaporation
P. Refrigerant
Q. Restriction
R. Tetrafluoroethane (R-134a)
S. Tubes
T. Vaporization

_____ **1.** The name given to a chemical compound designed to meet the needs of the refrigeration system.

_____ **2.** The process of transferring heat by the circulatory movement that occurs in a gas or fluid as areas of differing temperatures exchange places due to variations in density and the action of gravity.

_____ **3.** A system with a fixed orifice tube that uses an accumulator between the evaporator and the compressor.

_____ **4.** An air-conditioning door actuator controlled by electricity.

_____ **5.** Metal pipes running side to side or up and down that the coolant or refrigerant travels through.

_____ **6.** A totally self-contained system with no materials entering or exiting.

_____ **7.** The amount of heat required to change the state from a liquid to a gas without raising the actual gauge temperature.

_____ **8.** The changing of a gas into a liquid through cooling.

_____ **9.** The changing of a liquid to a gas through boiling.

_____ **10.** An inert, colorless gas that can be used as a refrigerant. It is stored in white containers.

_____ **11.** An inert colorless gas that can be used as a refrigerant. It is stored in light blue containers.

_____ **12.** The pressure of the air surrounding everything caused by gravity and the weight of air.

_____ **13.** A small, flat piece of metal placed between the tubes to help with the transfer of heat from the coolant to the air, refrigerant to air, or air to refrigerant.

_____ **14.** The flow of heat from a hotter part to a cooler part; it can occur in solids, liquids, or gases.

_____ **15.** A manufactured compound designed to be used as a refrigerant. It is now illegal due to the high chlorine content.

_____ **16.** The air-conditioning component used on fixed orifice systems to protect the compressor by storing liquid refrigerant so it does not reach the compressor.

_____ **17.** The air-conditioning component normally located in the passenger compartment designed to allow low-pressure refrigerant liquid to change states to a gas.

_____ **18.** A blockage that partially stops or slows the flow of a material such as refrigerant.

_____ **19.** The amount of heat removal necessary to change the state from a gas to a liquid without changing the actual gauge temperature.

_____ **20.** A measure of heat energy.

Multiple Choice

Read each item carefully, and then select the best response.

_____ 1. The federal law that regulates air emissions from stationary and mobile sources and contains all of the motor vehicle air-conditioning requirements and laws is called the _____.
   A. National Air Quality Standards Act
   B. Clean Air Act
   C. Air Conditioning Act
   D. Environmental Protection Act

_____ 2. The amount of water vapor in the air expressed as a percentage is called _____.
   A. saturation level
   B. condensation point
   C. relative humidity
   D. evaporation point

_____ 3. The process of transferring heat through matter by the movement of heat energy through solids from one particle to another is called _____.
   A. convection
   B. conduction
   C. radiation
   D. evaporation

_____ 4. Heat energy is measured in _____.
   A. degrees Fahrenheit
   B. degrees Celsius
   C. British thermal units
   D. Relative humidity

_____ 5. The heat required to change the water at its maximum temperature from a liquid to a gas is called _____.
   A. Latent heat of evaporation
   B. Latent heat of condensation
   C. Latent heat of freezing
   D. Latent heat of vaporization

_____ 6. The _____ changes a low-pressure refrigerant gas to a high-pressure gas and provides the needed refrigerant movement in the system.
   A. Condenser
   B. Compressor
   C. Evaporator
   D. Restrictor

_____ 7. Hot high-pressure gas enters the _____ from the compressor; the gas flows through a series of coils, ambient air removes the heat from the hot high-pressure gas, which begins to change into a liquid.
   A. Condenser
   B. Evaporator
   C. Restriction
   D. Orifice tube

_____ 8. Dichlorodifluoromethane, commonly known as _____, is a member of the CFC family of gases and was the first common refrigerant to be used in an automotive air conditioner.
   A. R-10
   B. R-12
   C. R-124b
   D. R-134a

_____ 9. All of the following are common oils used in refrigeration systems, _except_____.
   A. POE oil
   B. Mineral oil
   C. PAG oil
   D. Copaiba oil

_____ **10.** Fixed-orifice tube systems use a(n) _____ to ensure that a pure gas is delivered to the compressor.
   A. receiver filter drier
   B. accumulator
   C. schrader valve
   D. bernoulli filter

_____ **11.** The _____ is located inside the cabin in the plastic box with the air-conditioning evaporator.
   A. defroster
   B. condenser
   C. heater core
   D. accumulator

_____ **12.** Air doors are controlled by _____.
   A. cables
   B. vacuum servos
   C. electric servos
   D. All of the above

_____ **13.** The most common type of blower motor control is a(n) _____ that uses resistors in series to regulate the speed of the fan.
   A. resistor pack
   B. potentiometer
   C. resistor block
   D. either A or C

_____ **14.** The method of controlling a fan using "on" time compared to "off" time on the ground side of the circuit is known as _____ modulation.
   A. pulse width
   B. potential
   C. resistance
   D. current

_____ **15.** The term used to describe the standard air-conditioning testing process is _____ testing.
   A. condensation
   B. latent heat
   C. performance
   D. temperature

True/False

If you believe the statement to be more true than false, write the letter "T" in the space provided. If you believe the statement to be more false than true, write the letter "F".

_____ **1.** Automotive air-conditioning service and repair technicians need to have a special license.

_____ **2.** When pressure is lowered on a gas, its temperature rises; when pressure is raised, its temperature drops.

_____ **3.** The amount of heat given up by steam to turn back into a liquid is 970 Btu and is called latent heat of condensation.

_____ **4.** With the use of refrigerants, maintaining vaporization and condensing points at normal ambient temperatures is simply done by raising or reducing the pressure of the refrigerant.

_____ **5.** PAG oil mixed with mineral oil creates a hazardous gas that can corrode the air-conditioning system.

_____ **6.** The fixed orifice tube is adjustable based on the temperature of the outlet pipe from the evaporator, whereas the TXV provides a nonadjustable passage for the refrigerant to pass through.

_____ **7.** Many of the components used to warm the passenger cabin of a vehicle also function in removing heat from the engine.

_____ **8.** Attached to the spinning shaft of the blower motor is what is called the "bird cage," a plastic cage with fins to pull or push air.

_____ **9.** The heater core is a small radiator consisting of tubes, fins, and tanks.

_____ **10.** The compressor is the most common source of abnormal noises arising from the air conditioner.

Fill in the Blank

Read each item carefully, and then complete the statement by filling in the missing word(s).

1. Automotive air conditioning is federally regulated by the _____ _____ _____.

2. Heat always transfers from _____ objects to _____ objects.

3. The transfer of heat through the emission of energy in the form of invisible waves is called _____.

4. One pound of water at _____ _____ and at a temperature of 32°F (0°C) would require 1 Btu of energy added to increase the temperature 1 degree Fahrenheit.

5. At 32°F, the latent heat of _____ begins.

6. If the velocity of a liquid rises, the pressure of the liquid must drop according to the _____ principle.

7. Tetrafluoroethane, commonly known as _____, was the replacement for R-12 refrigerant.

8. Air _____ can be controlled by cables, vacuum servos, or electric servos.

9. Attached to the spinning shaft of the blower motor is a plastic cage with fins to pull or push air called a(n) _____ cage.

10. Some manufacturers use an activated charcoal cabin _____ _____, which helps trap odors and airborne pollutants such as carbon monoxide and oxides of nitrogen.

Labeling

Label the following diagrams with the correct terms.

1. Components of an air-conditioning unit:

A. _____

B. _____

C. _____

D. _____

E. _____

F. _____

G. _____

H. _____

I. _____

2. Heater controls:

Heater control valve:

A. _____

B. _____

C. _____

D. _____

E. _____

F. _____

G. _____

Blend door in the "Cold Air" position:

H. _____

I. _____

J. _____

K. _____

L. _____

M. _____

N. _____

Skill Drills

Test your knowledge of skill drills by filling in the correct words in the photo captions.

1. Performance Testing an Air-Conditioning System:

Step 1: Turn on the vehicle. Place a _____ in front of the vehicle to simulate the _____ that occurs when driving.

Step 2: Close all _____. Turn the air conditioner to its _____ setting.

Step 3: Raise the engine _____ to 1200–2000. Check the _____ temperature in the cabin using a _____. Compare the temperature recorded to the diagnostic _____ in the manufacturer's service manual.

2. Inspecting and Testing the HVAC Control Cables:

Step 1: Begin by _____ the operation and diagnostic procedure for the control panel assembly. Move the control panel _____ or _____, and observe the _____ on the air box for movement.

Step 2: If movement is not noted, _____ for a cable that is _____ or has popped off of the control _____ or door.

Step 3: If a _____ has popped off, reinstall it with the proper _____ and _____ the cable system.

3. Inspecting the Heater Ducts, Doors, and Cabin Filters:

Step 1: Start the vehicle and turn the _____ _____ to the _____ setting.

Step 2: Turn the _____ selector through all the positions and make sure the _____ is distributed through the appropriate _____. Turn the mode selector to _____ vents and _____.

Step 3: Take an average _____ measurement from all _____ (800 fpm minimum average). If no flowmeter is available, test the amount of airflow with your _____ or a piece of _____ and compare to a similar vehicle.

Step 4: Move the fresh air/recirculate control to the fresh air position and _____ the airflow to see if the cabin _____ is plugged. Make sure at least one _____ or door is open to avoid _____ airflow. The airflow should not drop.

Step 5: If the _____ air filter filters all air in the system, _____ the cabin air filter and hold it up to a _____ to see if it is plugged. _____ it if it is restricted or damaged.

Crossword Puzzle

Use the clues to complete the puzzle.

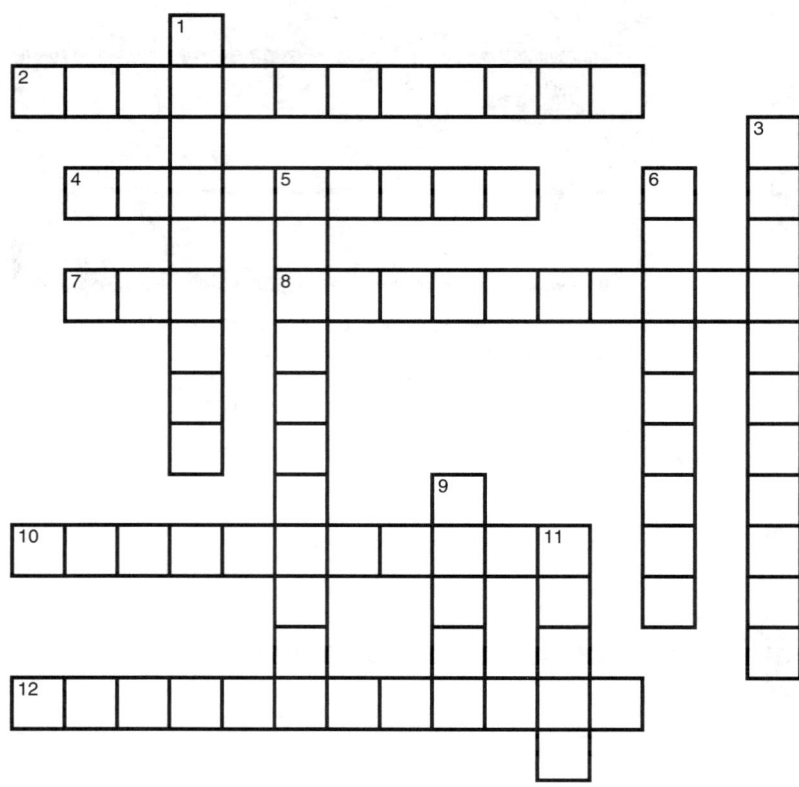

Across

2. The changing of a liquid to a gas through boiling.

4. The transfer of heat through the emission of energy in the form of invisible waves.

7. A small, flat piece of metal placed between the tubes to help with the transfer of heat from the coolant to the air, refrigerant to air, or air to refrigerant.

8. A belt- or electrically driven device designed to increase refrigerant pressure and cause refrigerant to travel through the air-conditioning system.

10. A policy signed into law in 1990 that sets standards for air pollution to eliminate ozone-depleting elements.

12. The flow of heat from a hotter part to a cooler part; it can occur in solids, liquids, or gases.

Down

1. The air-conditioning component located in the front of the vehicle designed to allow high-pressure refrigerant to change states from gas to liquid.

3. The name given to a chemical compound designed to meet the needs of the refrigeration system.

5. The air-conditioning component used on orifice systems to protect the compressor by storing liquid refrigerant so it does not reach the compressor.

6. A drying agent used in air-conditioning systems to absorb moisture.

9. Metal or plastic pieces that line the pipes used to connect the tubes together to allow the coolant to continue to flow.

11. Metal pipes running side to side or up and down that the coolant or refrigerant travels through.

ASE-Type Questions

Read each item carefully, and then select the best response.

_____ **1.** Tech A says that one advantage of an air conditioner is that the system removes water from the air. Tech B says that air conditioners are needed to remove moisture from the inside of the windshield to meet the mandated defrost time. Who is correct?
  A. Tech A
  B. Tech B
  C. Both A and B
  D. Neither A nor B

_____ **2.** Tech A says that the EPA regulates automotive air-conditioning systems. Tech B says that EPA tests air-conditioning systems every 5 years. Who is correct?
  A. Tech A
  B. Tech B
  C. Both A and B
  D. Neither A nor B

_____ **3.** Tech A says that an accumulator creates the pressure needed for the air-conditioning system to operate. Tech B says that the high side refers to refrigerant entering the compressor. Who is correct?
  A. Tech A
  B. Tech B
  C. Both A and B
  D. Neither A nor B

_____ **4.** Tech A says that heat is removed from the passenger compartment by the heater core. Tech B says that heat in the cab is removed through by the refrigerant and is released to the outside air at the condenser. Who is correct?
  A. Tech A
  B. Tech B
  C. Both A and B
  D. Neither A nor B

_____ **5.** Tech A says that conduction is when heat transfers through solids. Tech B says that radiation is when heat travels through space. Who is correct?
  A. Tech A
  B. Tech B
  C. Both A and B
  D. Neither A nor B

_____ **6.** Tech A says that the air-conditioning compressor creates high-pressure liquid. Tech B says that the metering device creates a physical change in refrigerant from a liquid to a gas. Who is correct?
  A. Tech A
  B. Tech B
  C. Both A and B
  D. Neither A nor B

_____ **7.** Tech A says that when an air-conditioning system freezes up, the refrigerant freezes and stops the cooling process. Tech B says that when an air-conditioning unit freezes up, the evaporator coils are covered with frozen water molecules that stop airflow across the coils, thus preventing cooling. Who is correct?
  A. Tech A
  B. Tech B
  C. Both A and B
  D. Neither A nor B

_____ **8.** Tech A says that the evaporator is on the high side of the system. Tech B says that the condenser transfers heat to the atmosphere. Who is correct?
  A. Tech A
  B. Tech B
  C. Both A and B
  D. Neither A nor B

_____ **9.** Tech A says that R-134a replaced R-12 as an approved refrigerant. Tech B says that HFO-1234yf refrigerant is flammable. Who is correct?

    A. Tech A

    B. Tech B

    C. Both A and B

    D. Neither A nor B

_____ **10.** Tech A says that the engine thermostat controls the available coolant temperature for the heater system. Tech B says that the heater core can have engine coolant or refrigerant flowing through it depending on whether heat or cool is commanded. Who is correct?

    A. Tech A

    B. Tech B

    C. Both A and B

    D. Neither A nor B

_____ **11.** Tech A states that an air-conditioning performance test usually requires that an auxiliary condenser fan be used during the test. Tech B states that a performance test will show if the air-conditioning system is contaminated with sealer. Who is correct?

    A. Tech A

    B. Tech B

    C. Both A and B

    D. Neither A nor B

Ignition Systems Overview

Chapter Review

The following activities have been designed to help you refresh your knowledge of this chapter. Your instructor may require you to complete some or all of these activities as a regular part of your training program. You are encouraged to complete any activity that your instructor does not assign as a way to enhance your learning.

Matching

Match the following terms with the correct description or example.

A. Advance mechanism
B. Breaker plate
C. Center electrode
D. Cranking
E. Direct ignition system
F. Distributor base plate
G. Electronic ignition system
H. Event cylinder
I. Heat range
J. Ignition coil

K. Ignition switch
L. Inductive current
M. Oscillate
N. Primary winding
O. Rotor
P. Secondary circuit
Q. Spark plug
R. Spark timing
S. Throttle body
T. Waste cylinder

_____ **1.** The low-voltage coiled copper wiring found in an ignition coil.

_____ **2.** The cylinder in a waste spark ignition system that receives a spark near the top of its exhaust stroke.

_____ **3.** A high-voltage rotating switch that transfers voltage from the distributor cap's center terminal to the outer terminals.

_____ **4.** The point at which a spark occurs at the spark plug relative to the position of the piston.

_____ **5.** A device used to amplify an input voltage into the much higher voltage needed to jump the electrodes of a spark plug.

_____ **6.** To cycle above and below a given value.

_____ **7.** The housing on an intake manifold that is used to control the amount of filtered air that enters the cylinders.

_____ **8.** The rating of a spark plug's operating temperature.

_____ **9.** A device used to trigger an earlier spark based on engine conditions.

_____ **10.** An ignition system that uses a nonmechanical method of triggering the ignition coil's primary circuit.

_____ **11.** The part of an ignition system that operates on higher voltage and delivers the necessary high voltage to the spark plugs.

_____ **12.** The movable plate the breaker points are mounted on that pivots as the vacuum advance pulls on it.

_____ **13.** May refer to a waste spark ignition or a coil-on-plug ignition system, in which the coils are directly attached to the spark plugs.

_____ **14.** The current that has been created across a conductor by moving it through a magnetic field.

_____ **15.** The electrode located in the center of a spark plug. It is the hottest part of the spark plug.

_____ **16.** A round metal plate near the top of the distributor that is attached to a distributor housing; also called a breaker plate.

_____ **17.** Rotating the engine by turning the ignition key to the start position.

_____ **18.** A switch operated by a key or start/stop button and used to turn on or off a vehicle's electrical and ignition system.

_____ **19.** The cylinder that uses the spark to ignite the air-fuel mixture on a waste spark ignition system.

_____ **20.** A device that provides a gap for the high-voltage spark to occur in each cylinder.

Multiple Choice

Read each item carefully, and then select the best response.

_____ **1.** For an engine to run smoothly and efficiently, the high-voltage spark must cross the spark plug electrode as the piston approaches top dead center of the _____.
A. intake stroke
B. compression stroke
C. power stroke
D. exhaust stroke

_____ **2.** What type of ignition system eliminates the distributor by using dedicated ignition coils—one coil for each pair of cylinders?
A. Contact breaker point ignition system
B. Electronic ignition system
C. Waste spark ignition system
D. Direct ignition system

_____ **3.** The _____ replaced the contact breaker points with an electronic switching device but still used a distributor to dispense the spark to the various cylinders.
A. electronic ignition system—distributor type
B. waste spark ignition system
C. direct ignition system
D. coil-on-plug ignition system

_____ **4.** The ignition system uses a(n) _____ to convert relatively low-voltage and high-current flow into very high-voltage and very low-current flow.
A. spark plug
B. ignition module
C. induction coil
D. reluctor

_____ **5.** Faraday's law states that relative movement between a conductor and a magnetic field allows _____ ways by which voltage can be induced in a conductor.
A. three
B. four
C. five
D. six

_____ **6.** The correct spark timing varies according to _____.
A. transmission gear selected
B. detected knock
C. engine temperature
D. all of the above

_____ **7.** The power train control module may also be called a(n) _____.
A. engine control module
B. body control module
C. electronic control unit
D. either A or C

_____ **8.** When the ignition switch is turned to the _____ position, most warning lamps on the instrument panel should illuminate.
A. accessory
B. on/run
C. start/crank
D. lock and off

_____ 9. When battery voltage pushes current through the _____ of an ignition coil, a magnetic field is created.
   A. primary winding
   B. secondary winding
   C. iron core
   D. high-tension terminal

_____ 10. Spark plugs are identified by _____.
   A. thread size or diameter
   B. reach or length of the thread
   C. heat range or operating temperature
   D. all of the above

_____ 11. Contact breaker points are opened by _____ on the distributor shaft, and closed by a spring.
   A. rubbing blocks
   B. sensors
   C. cam lobes
   D. gears

_____ 12. The _____ charges to the peak value of the primary winding voltage.
   A. rotor
   B. capacitor
   C. contact breaker
   D. condenser

_____ 13. It is the main function of the _____ to distribute the spark to the spark plugs in the correct sequence and at the correct time in the engine cycle.
   A. condenser
   B. ballast resistor
   C. distributor
   D. high-tension terminal

_____ 14. The vacuum advance mechanism is designed to operate on _____ vacuum.
   A. ported
   B. manifold
   C. venturi
   D. any of the above

_____ 15. A _____ is an example of an inductive pickup.
   A. Hall-effect switch
   B. throttle position sensor
   C. mass airflow sensor
   D. manifold absolute pressure sensor

_____ 16. The _____ has one tooth for each cylinder.
   A. reluctor
   B. conductor
   C. piston
   D. stator

_____ 17. The _____ is shaped like a very shallow cup with slits, or windows, cut into it at evenly spaced intervals and made of a ferrous metal.
   A. optical-type sensor
   B. interrupter ring
   C. reluctor
   D. stator

_____ 18. What type of sensor, located inside the distributor, can be used to sense the position of the crankshaft and send an appropriate voltage signal to the ignition module or PCM?
   A. Hall-effect
   B. Induction-type
   C. Optical-type
   D. Variable-resistance

_____ **19.** A distributorless ignition system uses a _____ to calculate engine speed and determine engine position.
  A. crankshaft position sensor
  B. cam lobe
  C. camshaft position sensor
  D. both A and C

_____ **20.** Some newer vehicles use high-performance spark plugs that are made with _____ and do not require replacement until the engine reaches 100,000 miles (160,000 km).
  A. platinum
  B. iridium
  C. gold palladium
  D. all of the above

True/False

If you believe the statement to be more true than false, write the letter "T" in the space provided. If you believe the statement to be more false than true, write the letter "F".

_____ **1.** The ignition coil's function is to amplify the battery's low voltage and high current into very high voltage and low current.

_____ **2.** The low-voltage side of the induction coil is called the secondary circuit, and the high-voltage side is called the primary circuit.

_____ **3.** Available voltage is the maximum amount of voltage available to try to push current to jump the spark plug gap if the gap were infinite.

_____ **4.** In early vehicles that used a distributor, base spark timing was set at idle speeds by positioning the distributor body in relation to its rotating cam.

_____ **5.** Ignition coils are basically step-up transformers.

_____ **6.** The secondary ignition coil winding has 15,000 to 30,000 turns of very thin insulated aluminum wire wound around its core.

_____ **7.** High-tension leads conduct the high-output voltage generated in the secondary ignition circuit between the high-tension terminal(s) of the ignition coil, the distributor cap, and the spark plugs.

_____ **8.** The same spark plug can sometimes be used in different engines with different gap settings.

_____ **9.** The condenser is a critical component of the primary circuit in point-type ignition systems. Without it, the points arc as they are opened.

_____ **10.** The high voltage induced in the secondary winding rises to a value great enough to bridge the gap across the spark plug.

_____ **11.** The distributor cap covers the end of the distributor to protect the components and provide a connection point between the rotor and the spark plug leads.

_____ **12.** The centrifugal advance mechanism controls ignition advance in relation to engine load.

_____ **13.** In electronic ignition systems, the contact breaker points are eliminated and the primary circuit is switched or triggered electronically with a power transistor located in the ignition module.

_____ **14.** The stator has one tooth for each cylinder; as it spins the teeth interact with the reluctor to trigger the ignition module.

_____ **15.** In a distributor, the Hall-effect generator and its integrated circuit are located on one leg of a U-shaped assembly, mounted on the distributor base plate.

_____ **16.** The optical sensor is very precise due to the small slits that the light passes through.

_____ **17.** In direct ignition systems, there is one ignition coil for each pair of companion cylinders.

_____ **18.** In a waste spark system, the cylinder on the compression stroke is said to be the event cylinder and the cylinder on the exhaust stroke is the waste cylinder.

_____ **19.** A spark test should be performed when an engine will not start, has one or more cylinders misfiring, or has other drivability problems, such as a lack of power, or stalling.

_____ **20.** Many technicians have been fooled by using one of the engine's spark plugs as a spark tester.

Fill in the Blank

Read each item carefully, and then complete the statement by filling in the missing word(s).

1. The purpose of the ignition system is to create the _____-_____ _____ and deliver it at the right time to each cylinder.

2. The original ignition system was called the _____ _____ _____ ignition system, a mechanical system with a switch that was opened and closed as the engine was running.

3. The amount of voltage required to initially get current to jump the spark plug gap is called _____ voltage.

4. Spark timing is almost always indexed to the _____ _____ cylinder.

5. The _____ _____ plate, also called a breaker plate, is a moveable metal plate located in the distributor, beneath the distributor cap on which the contact breaker points are mounted.

6. The _____ supplies the electrical energy to the ignition circuit during start-up.

7. The immobilizer system uses a randomly selected _____ _____, a constantly changing, numeric code that is communicated to the engine immobilizer and security system by a radio frequency signal.

8. _____-_____ _____ connect the secondary ignition components together, such as the coil to the distributor cap and the distributor cap to the spark plugs.

9. If the insulation around the cable is insufficient and the leads are not spaced far enough apart and run parallel to each other, then a(n) _____ _____ can be generated in the other wires.

10. Spark plug _____ is the distance from the seat of the spark plug to the end of the spark plug threads.

11. The amount of time, in degrees of _____ _____, that the contacts are closed is called the dwell angle.

12. The _____ advance mechanism controls ignition timing in relation to engine speed.

13. Ignition _____ are used in virtually every type of ignition system except the contact breaker point system.

14. _____-_____ systems are electronic systems that use a magnetic pulse generator, also called a variable reluctor sensor, to generate an AC signal.

15. The _____ has one tooth for each cylinder.

16. As the reluctor rotates, the teeth on the reluctor approach and leave the stator teeth, changing the _____ _____ between them.

17. In a distributor, the Hall-effect generator and its _____ _____ are located on one leg of a U-shaped assembly mounted on the distributor base plate.

18. In _____ ignition systems, also known as electronic ignition systems, the distributor is eliminated and replaced by multiple ignition coils.

19. While the _____ voltage is very high (up to 100,000V), the current flow is very low.

20. If the spark plug wires have a weak spot in the _____, you should be able to see and hear the spark jumping to another wire, the test lead, or conductive engine part.

Labeling

Label the following diagrams with the correct terms.

1. Contact breaker point ignition system:

A. _____

B. _____

C. _____

D. _____

E. _____

F. _____

G. _____

H. _____

I. _____

J. _____

K. _____

L. _____

M. _____

2. Electronic ignition system—distributor type:

A. _____

B. _____

C. _____

D. _____

E. _____

F. _____

G. _____

H. _____

I. _____

J. _____

K. _____

3. Waste spark ignition system:

A. _____

B. _____

C. _____

D. _____

E. _____

F. _____

G. _____

H. _____

I. _____

J. _____

4. Coil-on-plug ignition system:

A. _____

B. _____

C. _____

D. _____

E. _____

F. _____

G. _____

H. _____

I. _____

J. _____

5. Vacuum advance mechanism:

A. _____

B. _____

C. _____

D. _____

E. _____

F. _____

G. _____

H. _____

I. _____

J. _____

K. _____

L. _____

M. _____

N. _____

6. Ignition coil:

A. _____

B. _____

C. _____

D. _____

E. _____

F. _____

7. Parts of a high-tension lead:

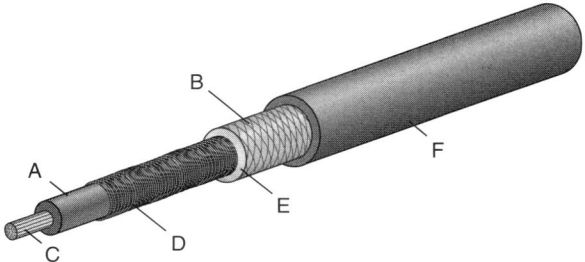

A. _____

B. _____

C. _____

D. _____

E. _____

F. _____

8. Spark plug:

A. _____

B. _____

C. _____

D. _____

E. _____

F. _____

G. _____

H. _____

I. _____

J. _____

9. Contact breaker point operation:

A. _____

B. _____

C. _____

D. _____

E. _____

F. _____

G. _____

H. _____

I. _____

J. _____

10. Distributor:

A. _____

B. _____

C. _____

D. _____

E. _____

F. _____

G. _____

H. _____

I. _____

J. _____

K. _____

L. _____

M. _____

N. _____

O. _____

P. _____

Q. _____

R. _____

S. _____

T. _____

U. _____

11. Vacuum sources:

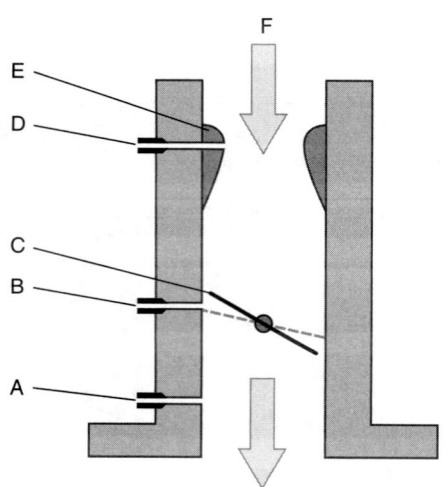

A. _____

B. _____

C. _____

D. _____

E. _____

F. _____

Skill Drills

Test your knowledge of skill drills by filling in the correct words in the photo captions.

1. Performing a Spark Test:

Step 1: Remove the high-tension _____ or _____ from the end of the _____ _____.

Step 2: Connect a _____ _____ to the _____ of the coil or high-tension lead, and attach the _____ to a good _____.

Step 3: Crank the engine, and watch for a _____ across the tester _____. A properly functioning _____ system will produce a _____ spark. If there is no spark present, test the spark plug _____ or coil, distributor cap, and _____ on that plug. If there is still no spark present, inspect the primary and secondary circuits.

2. Testing the Ignition Coil:

Step 1: Obtain the recommended testing procedure for the particular vehicle you are inspecting. Inspect and test ignition _____ and _____ windings of the ignition coil(s). All the coil _____ should be clean, secure, and free of _____.

Step 2: To test the _____ of the primary windings, place the _____ leads on each of the two primary winding _____. If the reading is not within specifications, the coil is _____ and must be _____.

Step 3: To test the _____ windings, place one ohmmeter _____ on the secondary _____ terminal and the other to one of the _____ terminals, or the coil secondary _____, if equipped. If the reading is not within specifications, the _____ is faulty and must be replaced.

Step 4: If the coil readings are OK, perform a _____ _____ to verify that the coil can produce a _____ that will _____ at least 1/2" (13 mm).

3. Testing a Spark Plug Wire:

Step 1: Disconnect the spark plug _____ at both ends by grasping the _____ on the ends and _____ while pulling the wire off. _____ each wire for _____, brittleness, or _____ spots.

Step 2: Place one lead of the _____ on each _____ of the spark plug wire. _____ the wire while _____ the ohmmeter.

Step 3: Replace the _____ _____ _____, start the engine, and lightly _____ water on the spark plug wires or run a grounded _____ _____ along each wire. If the readings are not within specifications or if the wires _____, replace the spark plug wires, making sure to _____ them in their factory positions to prevent damage.

Crossword Puzzle

Use the clues to complete the puzzle.

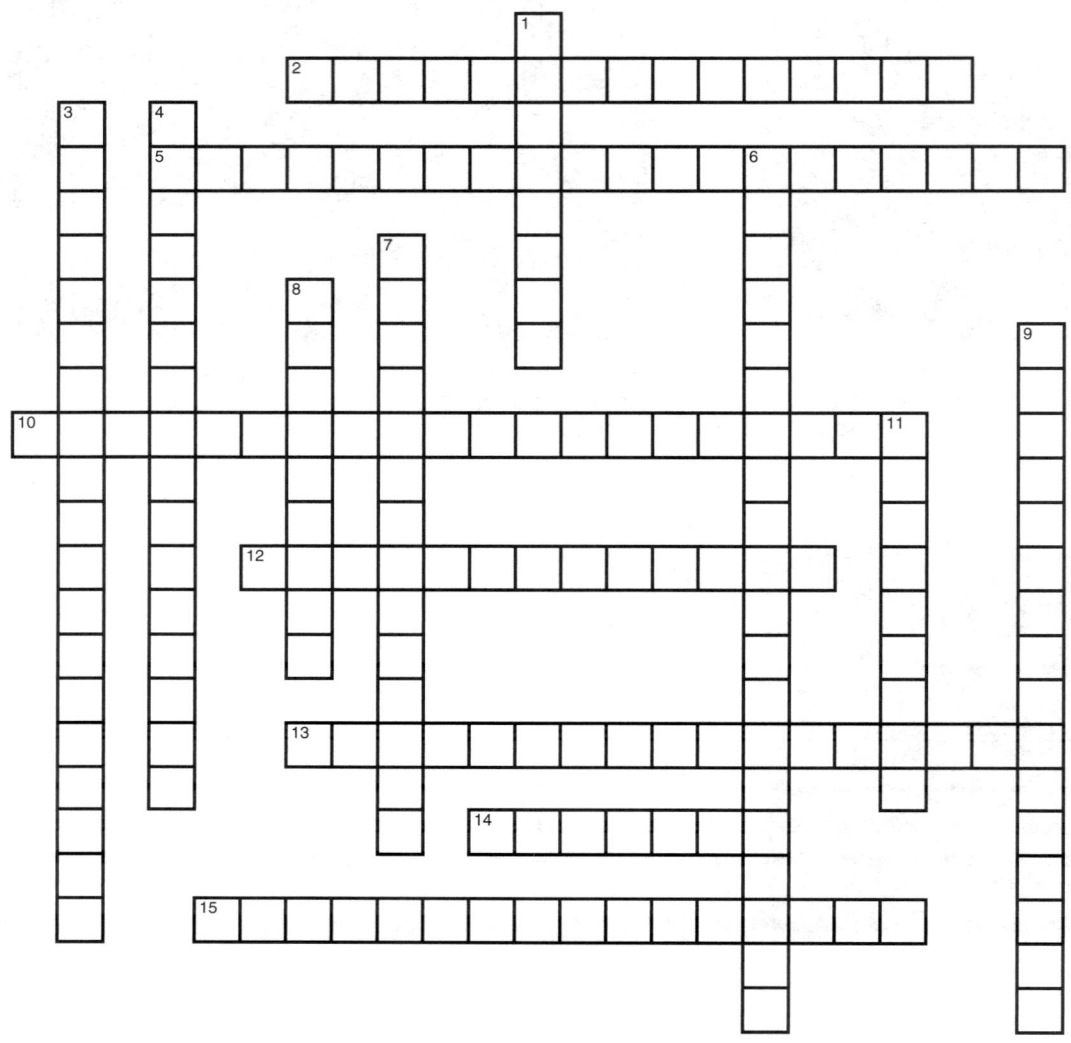

Across

2. A ferrous metal ring, shaped like a very shallow cup with slits or windows cut into it at evenly spaced intervals. The ring has the same number of blades and windows as engine cylinders and is rotated by the engine moving the blades through an air gap.

5. Pressure that is less than atmospheric pressure that develops in the intake manifold of a running engine equipped with a throttle plate.

10. A mechanically operated electrical switch that is fixed to the distributor base plate and opened and closed by the distributor cam with the rotation of the engine.

12. An electrical transformer that uses magnetic fields to produce high-voltage pulses from low-voltage direct current.

13. A mechanism that controls ignition timing advance in relation to engine load and causes spark at the spark plug to occur sooner based on engine conditions.

14. An electrochemical device used to supply voltage to a vehicle's electrical systems.

15. The maximum amount of voltage that the induction coil secondary is capable of putting out.

Down

1. Rotating the engine by turning the ignition key to the start position.
3. A computer that controls the ignition and fuel control and emissions control systems on an engine; also the electronic control unit (ECU) or power train control module (PCM).
4. The electrical arc that takes place between the center and the side electrode of a spark plug.
6. May refer to a waste spark ignition or a coil-on-plug ignition system, in which the coils are directly attached to the spark plugs.
7. The top portion of a distributor, used to make a connection between the spinning rotor and the high-tension leads.
8. The rating of a spark plug's operating temperature.
9. The portion of an electronic ignition system used to trigger the ignition system.
11. A device that provides a gap for the high-voltage spark to occur in each cylinder.

ASE-Type Questions

Read each item carefully, and then select the best response.

_____ 1. Tech A says that contact breaker points are a mechanical switch that opens and closes once for every ignition spark that is created. Tech B says that contact breaker points send high voltage directly from the points to the spark plugs. Who is correct?
  A. Tech A
  B. Tech B
  C. Both A and B
  D. Neither A nor B

_____ 2. Tech A says that as engines gain miles, the spark plug gap increases, which raises the ignition system's available voltage. Tech B says that misfire occurs when required voltage is higher than available voltage. Who is correct?
  A. Tech A
  B. Tech B
  C. Both A and B
  D. Neither A nor B

_____ 3. Tech A says that as engine rpm increases, spark timing generally increases. Tech B says that as engine load increases, spark timing generally decreases. Who is correct?
  A. Tech A
  B. Tech B
  C. Both A and B
  D. Neither A nor B

_____ 4. Tech A says that a Hall-effect switch uses light to turn a circuit on and off. Tech B says that waste spark systems don't need distributors. Who is correct?
  A. Tech A
  B. Tech B
  C. Both A and B
  D. Neither A nor B

_____ 5. Tech A says that coil-on-plug ignition systems use one coil to fire two cylinders. Tech B says that you should twist spark plug boots before removing them. Who is correct?
  A. Tech A
  B. Tech B
  C. Both A and B
  D. Neither A nor B

_____ 6. Tech A says that the ignition system will maintain spark at the spark plug for approximately 23 degrees of crankshaft rotation. Tech B says that the duration of spark in the spark plug only lasts 2 to 3 degrees of crankshaft rotation. Who is correct?
  A. Tech A
  B. Tech B
  C. Both A and B
  D. Neither A nor B

_____ **7.** Tech A says that the positive side of the coil primary circuit is typically switched by the module. Tech B says that the negative side of the coil primary circuit is typically switched by the module. Who is correct?
   A. Tech A
   B. Tech B
   C. Both A and B
   D. Neither A nor B

_____ **8.** Tech A says that one main advantage of distributorless ignition systems is no moving parts to maintain. Tech B says that the coil-on-plug ignition system uses a rotor to distribute the spark to each cylinder. Who is correct?
   A. Tech A
   B. Tech B
   C. Both A and B
   D. Neither A nor B

_____ **9.** When removing spark plugs, Tech A says that you should blow around each spark plug to remove any debris that could fall into the cylinder. Tech B says that on engines that have more than about 100,000 miles, you should adjust the spark plug gap to about half the specified gap. Who is correct?
   A. Tech A
   B. Tech B
   C. Both A and B
   D. Neither A nor B

_____ **10.** Tech A says that dielectric grease applied to the inside of the spark plug boot is used to keep the boot from sticking to the spark plug porcelain. Tech B says that the spark plug boots may need to be "burped" so that the air doesn't push the boot off the spark plug. Who is correct?
   A. Tech A
   B. Tech B
   C. Both A and B
   D. Neither A nor B

Gasoline Fuel Systems

Chapter Review

The following activities have been designed to help you refresh your knowledge of this chapter. Your instructor may require you to complete some or all of these activities as a regular part of your training program. You are encouraged to complete any activity that your instructor does not assign as a way to enhance your learning.

Matching

Match the following terms with the correct description or example.

A. Air supply system
B. Dieseling
C. Element
D. Fuel system
E. Idle

F. Knocking
G. Pressure
H. Throttle
I. Vacuum
J. Vapor lock

_____ 1. A condition in which the engine continues to run after the ignition key is turned off. Also referred to as run-on.

_____ 2. The speed at which an engine runs without any throttle applied.

_____ 3. The force per unit area applied to the surface of an object.

_____ 4. A device used to produce acceleration by controlling the air/fuel mixture.

_____ 5. A situation in which vapor forms in the fuel line, and the bubbles of vapor block the flow of fuel and stop the engine.

_____ 6. Equipment in a motor vehicle that delivers fuel to the engine.

_____ 7. The replaceable portion of a filter, such as an air filter or oil filter.

_____ 8. A pressure in an enclosed area that is lower than atmospheric pressure.

_____ 9. A noise heard when the air/fuel mixture spontaneously ignites before the spark plug is fired at the optimum ignition moment.

_____ 10. Equipment in a motor vehicle that delivers air to the engine.

Multiple Choice

Read each item carefully, and then select the best response.

_____ 1. If liquid gasoline vaporizes in the fuel pump, bubbles of vapor can block the flow of fuel and stop the engine causing _____.
A. engine knock
B. dieseling
C. choking
D. vapor lock

_____ 2. A violent collision of flame fronts in the cylinder, caused by uncontrolled combustion is called _____.
A. dieseling
B. detonation
C. misfire
D. throttling

_____ **3.** The stoichiometric air/fuel ratio is represented by the Greek letter _____.
   A. alpha
   B. delta
   C. omega
   D. lambda

_____ **4.** All of the following are key components of internal combustion, *except*: _____.
   A. air
   B. fuel
   C. heat
   D. pressure/vacuum differential

_____ **5.** Atmospheric pressure at sea level is calculated as _____.
   A. 14.7 psi
   B. 1 lambda
   C. 101 kilopascals
   D. either A or C

_____ **6.** Vapor from the fuel tank is vented through a _____ where fuel vapors are stored until they are burned in the engine.
   A. drier
   B. charcoal canister
   C. desiccant
   D. filter sock

_____ **7.** The _____ is a variable resistor that is attached to a float mechanism in the fuel tank.
   A. sending unit
   B. fuel pump relay
   C. pressure regulator
   D. accelerometer

_____ **8.** A _____ is incorporated into the end of the fuel pickup tube and is the first line of defense against fuel contamination.
   A. fuel filter
   B. fuel rail
   C. filter sock
   D. fuel separator

_____ **9.** The _____ system uses one or two fuel injectors located centrally on the intake manifold, right above the throttle plates.
   A. throttle body injection
   B. fuel rail injection
   C. diesel fuel direct injection
   D. gasoline direct injection

_____ **10.** The _____ system uses a fuel injector for each cylinder located in the intake manifold near each intake valve that sprays fuel toward the valve.
   A. single-point injection
   B. multipoint fuel injection
   C. gasoline direct injection
   D. throttle body injection

True/False

If you believe the statement to be more true than false, write the letter "T" in the space provided. If you believe the statement to be more false than true, write the letter "F".

_____ **1.** The more effectively liquid gasoline is changed into vapor, the more efficiently it burns in the engine.

_____ **2.** There are two different methods used to measure the octane rating of a fuel—the Research Octane Number and the Motor Octane Number.

_____ **3.** A lean air/fuel mixture has less air in proportion to the amount of fuel.

_____ **4.** Dieseling may cause an engine to run backward for a brief time when it comes to a stop.

_____ **5.** A vacuum gauge can be calibrated in inches of mercury in a scale reading from 0" to 30", or if using millimeters of mercury, the scale reads from 0 to 760 mm.

_____ **6.** Modern vehicles are required by the Environmental Protection Agency to have a vented gas cap.

_____ **7.** The fuel filter typically consists of a pleated paper filter housed in a sealed container, and its primary function is to prevent contaminants from reaching the injectors.

_____ **8.** Some vehicles use a pressure regulator on the fuel rail, some are in the fuel tank, and others control fuel pressure by controlling the speed of the fuel pump.

_____ **9.** Gasoline direct injection engines can run fuel mixtures as lean as 65 to 1, much leaner than the stoichiometric ratio.

_____ **10.** Sequential injection means that the injectors operate twice per cycle, once each crankshaft revolution, each time delivering half the fuel for the cycle.

Fill in the Blank

Read each item carefully, and then complete the statement by filling in the missing word(s).

1. The _____ _____ system draws in gasoline from the gas tank (fuel cell) and delivers it under pressure to a fuel metering device.

2. The less easily the fuel ignites, the higher the _____ _____.

3. The term _____ _____ describes the chemically correct air/fuel ratio necessary to achieve complete combustion of the fuel and air.

4. _____ is one of the essential components of the internal combustion engine.

5. At _____ speed, the pistons draw air away from the manifold at a faster rate than it can pass the throttle plate into the manifold. creating a high vacuum, or low pressure.

6. The fuel _____ _____ can incorporate the use of a blowback ball valve to prevent fuel from leaking from the vehicle during fill-ups and to deter gas theft.

7. The primary job of the _____ _____ is to send constant electrical signals to the gas gauge located in the driver information center or to the BCM, which then controls the gas gauge.

8. A(n) _____ _____ is a special manifold designed to provide a reservoir of pressurized fuel for the fuel injectors.

9. The modern _____ _____ is simply a spring-loaded, electric-solenoid spray nozzle.

10. There are two types of _____ fuel injection systems. One uses a pressure regulator in the fuel tank, and the other controls the speed of the fuel pump to modify pressure.

Labeling

Label the following diagrams with the correct terms.

1. Fractional distillation tower:

A. _____

B. _____

C. _____

D. _____

E. _____

F. _____

G. _____

H. _____

I. _____

J. _____

2. Fuel tank:

A. _____

B. _____

C. _____

D. _____

E. _____

F. _____

G. _____

H. _____

I. _____

J. _____

Crossword Puzzle

Use the clues to complete the puzzle.

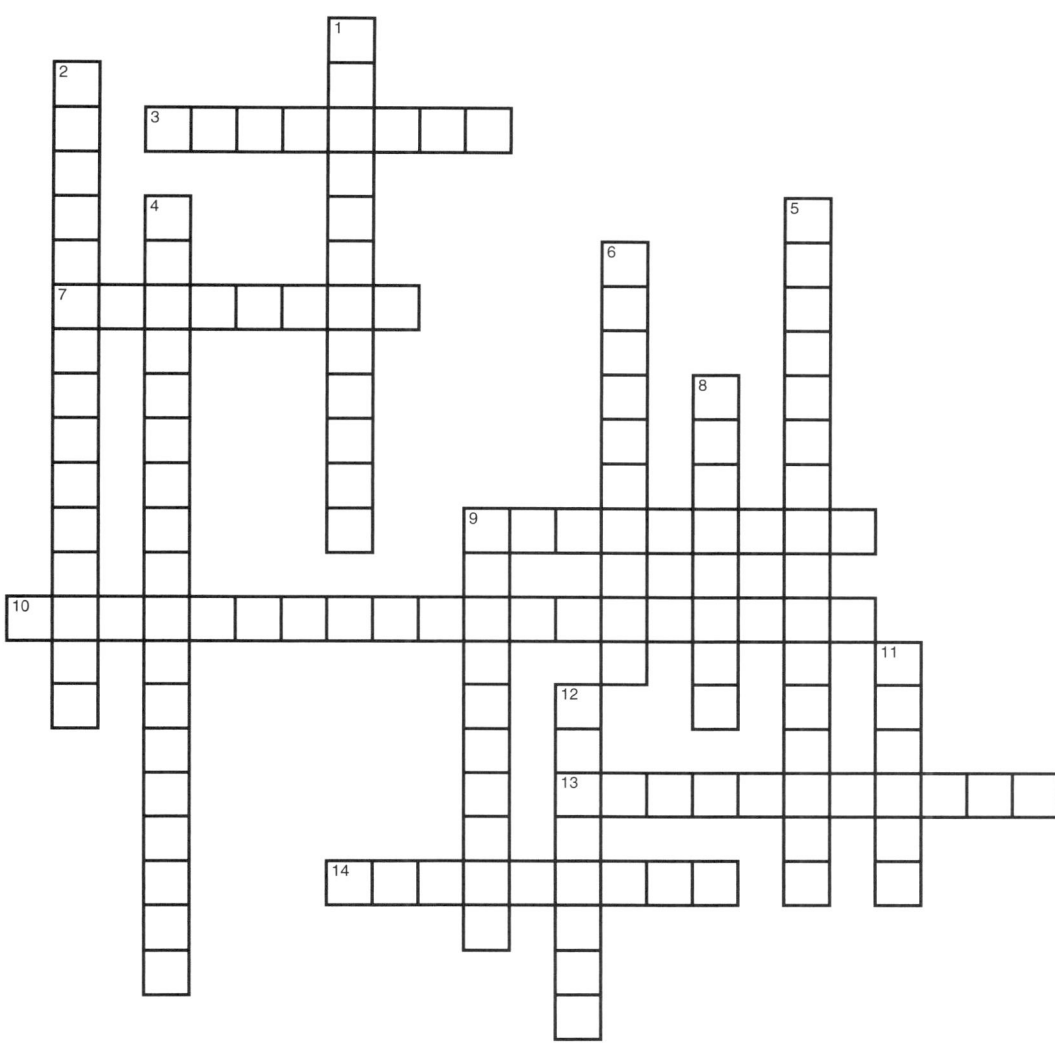

Across

3. A device used to produce acceleration by controlling the air-fuel mixture.

7. The force per unit area to the surface of an object.

9. A condition in which the engine continues to run after the ignition key is turned off. Also referred to as run-on.

10. The optimum ratio of air to fuel for combustion--14.7 parts air to 1 part fuel by weight.

13. The component in the fuel supply system responsible for sending constant electrical signals to the gas gauge located in the driver information center.

14. A situation in which vapor forms in the fuel line, and the bubbles of vapor block the flow of fuel and stop the engine.

Down

1. A standard measure of the performance of a motor or aviation fuel.
2. Equipment in a motor vehicle that delivers air to the engine.
4. Equipment in a motor vehicle that delivers the proper amount of fuel entering the injectors.
5. A device used for transmitting a signal to control a fuel gauge.
6. The first line of defense in the fuel supply system. The sock typically consists of a fine mesh, which prevents most small particles from being drawn into the fuel pump and sent through the rest of the fuel system.
8. Tubing that connects several injectors to the main fuel line.
9. The condition in which the remaining fuel charge fires or burns too rapidly after the initial combustion of the air-fuel mixture. It is audible through the combustion chamber walls as a knocking noise.
11. A pressure in an enclosed area that is lower than atmospheric pressure.
12. A volatile, flammable liquid mixture of hydrocarbons, obtained from crude oil and used as fuel for internal combustion engines.

ASE-Type Questions

Read each item carefully, and then select the best response.

_____ 1. Tech A says that most electric fuel pumps are now mounted inside of the fuel tank. Tech B says that fuel flows through the center of the electric pump and is used to cool the pump. Who is correct?
   A. Tech A
   B. Tech B
   C. Both A and B
   D. Neither A nor B

_____ 2. Tech A says that the stoichiometric ratio is 14.7 gallons of air to 1 gallon of fuel. Tech B says it is 14.7 lb. of air to 1 lb. of fuel. Who is correct?
   A. Tech A
   B. Tech B
   C. Both A and B
   D. Neither A nor B

_____ 3. Tech A says that some throttle plates are not mechanically connected to the gas pedal, but are operated by an electric motor. Tech B says that too much alcohol in gasoline can cause the engine to not run properly. Who is correct?
   A. Tech A
   B. Tech B
   C. Both A and B
   D. Neither A nor B

_____ 4. Tech A says that high octane fuel is harder to ignite than lower octane. Tech B says that using high octane fuel in an engine designed for lower octane will produce better fuel economy and power. Who is correct?
   A. Tech A
   B. Tech B
   C. Both A and B
   D. Neither A nor B

_____ 5. Tech A says that dieseling in a car today indicates a possible leaking injector. Tech B says that a tripped inertia switch means that there is a possible short in the fuel pump circuit. Who is correct?
   A. Tech A
   B. Tech B
   C. Both A and B
   D. Neither A nor B

_____ 6. Tech A says that a lambda greater than 1 means the engine is running rich. Tech B says that the amount of oxygen in the exhaust indicates how rich or lean the air/fuel mixture is. Who is correct?
   A. Tech A
   B. Tech B
   C. Both A and B
   D. Neither A nor B

_____ **7.** Tech A says that gauge pressure indicates pressure above atmospheric pressure. Tech B says that atmospheric pressure increases as elevation increases. Who is correct?
- **A.** Tech A
- **B.** Tech B
- **C.** Both A and B
- **D.** Neither A nor B

_____ **8.** Tech A says that fuel pressure is regulated by a fuel pressure regulator on some vehicles. Tech B says that fuel pressure is regulated by controlling the speed of the fuel pump on some vehicles. Who is correct?
- **A.** Tech A
- **B.** Tech B
- **C.** Both A and B
- **D.** Neither A nor B

_____ **9.** Tech A says that measuring the alcohol content in gasoline involves using water. Tech B says that as long as the fuel pressure is correct, you don't have to worry about fuel pump volume. Who is correct?
- **A.** Tech A
- **B.** Tech B
- **C.** Both A and B
- **D.** Neither A nor B

_____ **10.** When testing fuel injectors, Tech A says that the resistance can be tested and compared to specifications. Tech B says that resistance can be tested by measuring the fuel pressure drop when activating them with an injector pulse tool. Who is correct?
- **A.** Tech A
- **B.** Tech B
- **C.** Both A and B
- **D.** Neither A nor B

On-Board Diagnostics

Tire Tread:
© AbleStock

Chapter Review

The following activities have been designed to help you refresh your knowledge of this chapter. Your instructor may require you to complete some or all of these activities as a regular part of your training program. You are encouraged to complete any activity that your instructor does not assign as a way to enhance your learning.

Matching

Match the following terms with the correct description or example.

A. Aftermarket

B. Bidirectional scanners

C. Body control module (BCM)

D. Clean Air Act

E. Continuous monitoring

F. Controller area network (CAN)

G. Data link connector (DLC)

H. Diagnostic trouble code (DTC)

I. Drive cycle

J. Electronic brake control module (EBCM)

K. Emissions

L. Emission analyzer

M. Freeze-frame

N. History code

O. Malfunction indicator lamp (MIL)

P. Module

Q. Monitor

R. Power train control module (PCM)

S. Scan tool

T. Transmission control module (TCM)

_____ **1.** Tailpipe and volatile organic compound pollutants emitted by the automobile.

_____ **2.** An OBDII test run to ensure that a specific component or system is working properly.

_____ **3.** A series of prescribed automobile operating conditions during which emissions testing is performed.

_____ **4.** A plug-in electronic device for extracting and interpreting fault codes (DTCs) in the automobile, plus much more.

_____ **5.** A localized (on-board) vehicle network that enables computers and components to send and receive signals across a shielded twisted pair of wires.

_____ **6.** An electronic computer that controls transmission function; it may include adaptive learning capabilities for driver preferences.

_____ **7.** A computer-driven on-board report regarding component or system faults following the running of monitors.

_____ **8.** An electronic computer or circuit board that controls specific functions.

_____ **9.** The module that controls and monitors the anti-lock braking system.

_____ **10.** Federal legislation enacted in the United States to reduce pollution, in part from automobiles.

_____ **11.** A feature of OBDII that records events before, during, and after a fault occurs.

_____ **12.** Scanners used to monitor engine compression, vacuum, internal engine anomalies, and so forth by causing various components and systems to operate for test purposes.

_____ **13.** The module that controls and monitors the engine ignition, fuel, and emission system functions.

_____ **14.** A term that describes OBDII monitors that run continuously throughout the drive cycle.

_____ **15.** That segment of the trade that supplies parts, services, and repair for vehicles outside of the original equipment manufacturer (OEM) or the dealer network.

_____ **16.** A fault code that has occurred but is not current and is saved in the PCM's memory for 40 drive cycles.

_____ **17.** A device that enables a scan tool to access data stored in the vehicle's various computers.

_____ **18.** A service bay or lab device used for detecting/measuring vehicle emissions.

_____ **19.** An electronic unit that monitors and regulates electronic devices in the vehicle.

_____ **20.** An indicator located in the instrument cluster that illuminates when the power train control module (PCM) detects a fault in one of the vehicle systems. Formerly called a check engine or service engine soon light.

Multiple Choice

Read each item carefully, and then select the best response.

_____ **1.** When a vehicle's _____ is illuminated it means the vehicle is not complying with clean air regulations.
   A. malfunction indicator lamp
   B. VOC indicator light
   C. check engine light
   D. emission indicator lamp

_____ **2.** Which of the following pollutants are monitored and controlled by on-board systems using state-of-the-art electronics?
   A. Hydrocarbons
   B. Carbon dioxide
   C. Oxides of nitrogen
   D. All of the above

_____ **3.** Oxidizing catalytic converters were introduced in _____.
   A. 1968
   B. 1971
   C. 1975
   D. 1981

_____ **4.** The second generation of on-board diagnostic (OBDII) systems operates under standards set by _____.
   A. Association des Constructeurs Européens d'Automobiles
   B. Society of Automotive Engineers
   C. Japanese Automotive Standards Organization
   D. all of the above

_____ **5.** The complete listing of OBDI and OBDII standardized nomenclature for parts and systems used by engineers and technicians is known as _____.
   A. SAE J2012
   B. SAE J1930
   C. SAE J1962
   D. SAE J1968

_____ **6.** To simplify wiring, _____ have become common place in today's vehicles.
   A. controller area networks
   B. data link connectors
   C. OBDI systems
   D. closed-loop systems

_____ **7.** All diagnostic trouble codes have been standardized by the Society of Automotive Engineers and are listed in _____.
   A. SAE J1972
   B. SAE J1930
   C. SAE J2012
   D. SAE J1962

_____ **8.** If the first character of an OBDII diagnostic code is the letter _U_ then the fault is located in the _____.
   A. body
   B. communication system
   C. chassis controller
   D. emission system

_____ 9. Snapshots that are automatically recorded in the vehicle's power train control module when a vehicle fault occurs are referred to as _____ data.
   A. freeze-frame
   B. continuously monitored
   C. keep alive
   D. OBDI

_____ 10. A _____ includes the following events: A vehicle starts, warms up, is accelerated, cruises, slows down, accelerates once more, decelerates, stops, and cools down.
   A. warm-up cycle
   B. scan cycle
   C. drive cycle
   D. fault cycle

True/False

If you believe the statement to be more true than false, write the letter "T" in the space provided. If you believe the statement to be more false than true, write the letter "F".

_____ 1. Carbon monoxide and sulfur dioxide react together to create ground-level ozone, which is considered a health hazard.

_____ 2. Evaporative emission control systems were first introduced in 1971.

_____ 3. OBDI monitored mainly for parts and wiring malfunctions. OBDII is an enhanced diagnostic system that identifies faults in anything that may affect the vehicle's emission system.

_____ 4. Emission-related diagnostic trouble codes are the same across all vehicle makes and models, as are the SAE-recommended names used to describe components and systems.

_____ 5. If the catalytic converter is at risk, such as from overfueling or a continuous misfire, the malfunction indicator lamp will flash.

_____ 6. Diagnostic trouble codes are saved even if the vehicle's battery is disconnected.

_____ 7. OBDI systems monitor all emission-related components and circuits for opens, shorts, and abnormal operation.

_____ 8. There are hundreds of possible diagnostic trouble codes, and the list grows constantly as on-board systems become more sophisticated and unique.

_____ 9. An evaporative emission monitor will not run if the fuel tank is nearly full or nearly empty.

_____ 10. The bidirectional scan tool is like a "magic bullet." The fault code points directly to the problem.

Fill in the Blank

Read each item carefully, and then complete the statement by filling in the missing word(s).

1. _____ _____ compounds may be fuel or oil vapors emitted from the fuel tank, fuel lines, engine crankcase, or elsewhere.

2. OBDII faults and background data can be accessed and read by original equipment manufacturer (OEM) test equipment or by _____ test equipment made by other companies.

3. The first generation of _____-_____ diagnostic systems operated under manufacturer standards, starting with California vehicles and becoming nationwide in 1981.

4. The _____ _____ _____ is a 16-pin connector with a common (SAE J1962) size and shape.

5. Diagnostic trouble _____ are "set" in one or more of the vehicle's on-board computers once a fault is detected.

6. Possible faults, including engine misfiring and an incorrect air/fuel mixture, are monitored on a continuous basis by the _____ _____ control module.

7. Lower-priority faults are monitored on a _____ basis, meaning they are checked only once during each engine warm-up cycle, or even less often, depending on certain circumstances such as ambient temperatures or fuel level.

8. A diagnostic trouble code will remain logged in the PCM's memory as a(n) _____ _____ for a set period.

9. A(n) _____ _____ is a device able to electronically communicate with and extract data from the vehicle's one or more on-board computers.

10. More sophisticated than simple scan tools, _____ scanners are used to cause various components and systems to operate for test purposes.

Skill Drills

Test your knowledge of skill drills by filling in the correct words in the photo captions.

1. Retrieving and Recording DTCs, OBD Monitor Status, and Freeze-Frame Data:

Step 1: Select the _____ _____ to provide the best coverage for the _____ and _____ of vehicle. Locate the _____, and connect the scan tool.

Step 2: Power on the scan tool, and turn the _____ on. Establish scan tool _____ with the _____.

Step 3: _____ and _____ the DTCs.

Step 4: Retrieve and record _____ _____ status.

Step 5: Retrieve and record _____-_____ data applicable to DTCs and _____.

Step 6: Power _____ the scan tool, turn the _____ off, and _____ the scan tool.

Crossword Puzzle

Use the clues to complete the puzzle.

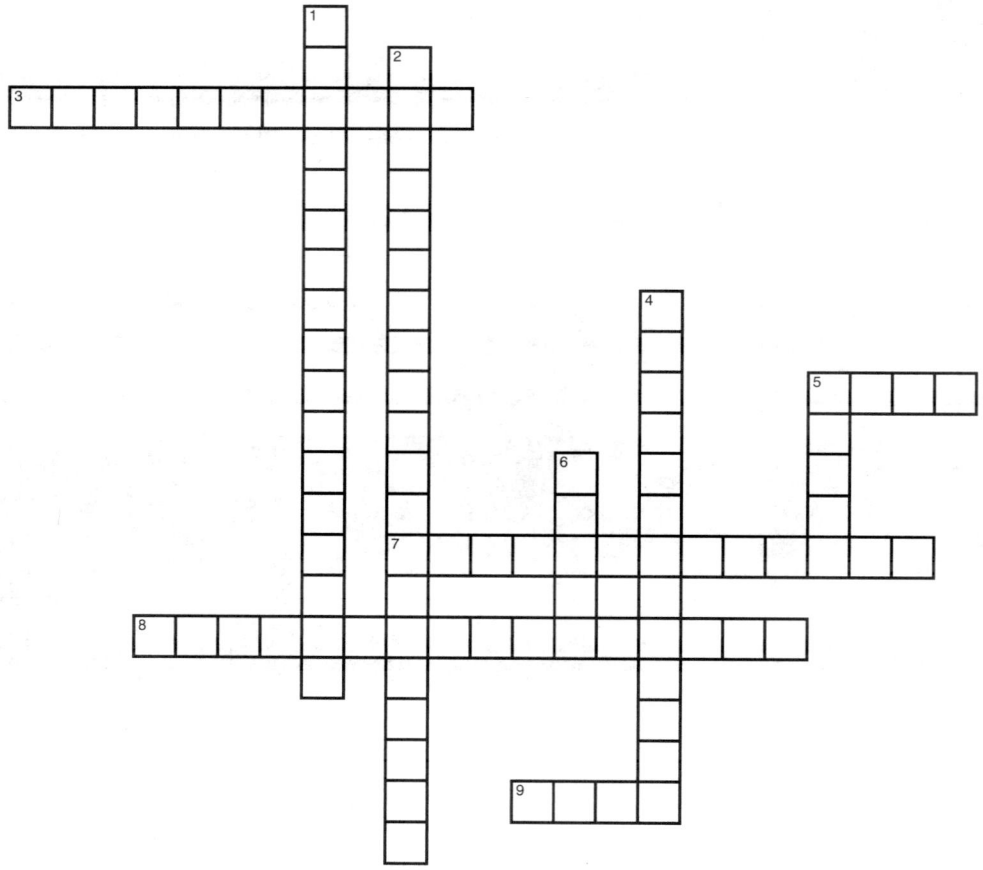

Across

3. A microscopic unburned fuel particle that contributes to photochemical smog and helps form ground-level ozone.

5. The first generation of on-board diagnostic systems that originated for California vehicles.

7. A pollutant resulting from sulfur in motor fuel and contributing to acid rain.

8. A vehicle emission that contributes to ground-level ozone. Oxides of nitrogen are produced when nitrogen and oxygen react during combustion, given sufficient temperatures and pressures.

9. Another name for a fault code.

Down

1. Unseen microscopic particles of carbon consisting of soot (especially prominent with diesel exhaust). Also known as PM, this clogs the lungs and is carcinogenic.

2. A monitor that runs only once per drive cycle.

4. A vehicle emission that is considered a primary greenhouse gas, though not toxic, and not yet regulated.

5. The second generation of on-board diagnostic systems, which have been in effect for all US vehicles since 1996.

6. Another word for diagnostic trouble, or a computer-driven on-board report regarding component or system faults following the running of monitors.

ASE-Type Questions

Read each item carefully, and then select the best response.

_____ **1.** Tech A says that on OBDII vehicles, it is a good idea to clear the codes before diagnosis, and see if they reset. Tech B says that the DTC will tell you what part needs to be changed. Who is correct?
  A. Tech A
  B. Tech B
  C. Both A and B
  D. Neither A nor B

_____ **2.** Tech A says that OBDI and OBDII use different DLC connectors. Tech B says that OBDII standardizes the designations for diagnostic trouble codes (DTCs). Who is correct?
  A. Tech A
  B. Tech B
  C. Both A and B
  D. Neither A nor B

_____ **3.** Tech A says that monitors are designed to test if emission systems are working properly. Tech B says that monitors are designed to store sensor data about a fault if a DTC is set. Who is correct?
  A. Tech A
  B. Tech B
  C. Both A and B
  D. Neither A nor B

_____ **4.** Tech A says that control modules communicate back and forth today using a CAN, which is a bundle of many individual wires, specially designed for fast communication speed. Tech B says that OBDII mandates that a DTC must be set if the emissions are more than 1.5 times the EPA FTP. Who is correct?
  A. Tech A
  B. Tech B
  C. Both A and B
  D. Neither A nor B

_____ **5.** Tech A says that OBDII will allow a technician to hook up a generic scan tool to read DTCs and clear DTCs. Tech B says that when the MIL is on, the vehicle should not be driven. Who is correct?
  A. Tech A
  B. Tech B
  C. Both A and B
  D. Neither A nor B

_____ **6.** Tech A says that something as simple as a loose gas tank fill cap will turn on the MIL. Tech B says that every digit of a DTC has identifying characteristics. Who is correct?
  A. Tech A
  B. Tech B
  C. Both A and B
  D. Neither A nor B

_____ **7.** Tech A says that a "P" in the DTC stands for a powertrain code. Tech B says a "U" stands for an undetermined code. Who is correct?
   A. Tech A
   B. Tech B
   C. Both A and B
   D. Neither A nor B

_____ **8.** Tech A says that on OBDII systems, the KAM stores DTCs forever. Tech B says that on OBDII systems, the KAM is erased when the battery is disconnected. Who is correct?
   A. Tech A
   B. Tech B
   C. Both A and B
   D. Neither A nor B

_____ **9.** Tech A says that the MIL can turn off if the fault doesn't reappear in a certain number of tests. Tech B says that when an MIL is activated, the PCM stores the DTC until a predetermined number of drive cycles have been performed or cleared by a scan tool. Who is correct?
   A. Tech A
   B. Tech B
   C. Both A and B
   D. Neither A nor B

_____ **10.** Tech A says some mechanical engine problems can cause OBDII DTCs to be set. Tech B says that OBDII codes only monitor the emission control system components. Who is correct?
   A. Tech A
   B. Tech B
   C. Both A and B
   D. Neither A nor B

Induction and Exhaust

Chapter Review

The following activities have been designed to help you refresh your knowledge of this chapter. Your instructor may require you to complete some or all of these activities as a regular part of your training program. You are encouraged to complete any activity that your instructor does not assign as a way to enhance your learning.

Matching

Match the following terms with the correct description or example.

A. Blow-off valve

B. Compressor surge

C. Forced induction

D. Helmholtz resonator

E. Mandrel forming

F. Mass airflow (MAF) sensor

G. Plenum chamber

_____ **1.** The pressurization of airflow going into the cylinder through the use of a turbocharger or supercharger.

_____ **2.** The special bending of pipe to ensure the pipe does not collapse. The use of this pipe bender allows for very tight bends without creating kinks or reducing the size of the pipe.

_____ **3.** The PCM input sensor that tells the computer the amount of air coming into the engine. The sensor allows the correct calculated amount of fuel delivery to the engine. It is usually located in the ducting of the air cleaner system.

_____ **4.** The backup of air against the throttle plate as it is closed. The turbocharger is still spinning, pressurizing air, when the throttle plate is closed. Air will stack up, creating a rapid slowing of the turbocharger compressor wheel. This can damage the compressor wheel.

_____ **5.** A valve that allows the release of excessive boost pressure from the turbocharger when the throttle plate is quickly closed.

_____ **6.** A device that uses the principle of noise cancellation through the collision of sound waves. When necessary, it is used in addition to the muffler to cancel additional sounds. This device may also be used on the induction system to muffle noise of airflow through the induction system.

_____ **7.** A large portion of the intake manifold after the throttle plate and before the intake runner tubes. The manifold provides a reservoir of air and helps prevent interference with the flow of air between individual branches.

Multiple Choice

Read each item carefully, and then select the best response.

_____ **1.** The air cleaner element or filter is manufactured using _____.

    A. pleated paper

    B. oil-impregnated cloth or felt

    C. an oil bath configuration

    D. any of the above

_____ **2.** A(n) _____ is a container that is sealed and specially shaped to cancel noise created by pressure waves.

    A. air cleaner

    B. baffle

    C. Helmholtz resonator

    D. intake manifold

_____ 3. Cylinder heads that have intake and exhaust manifolds on opposite sides of the engine are known as _____ heads.
   A. parallel
   B. cross-flow
   C. in-line
   D. variable-intake

_____ 4. Manifolds that respond to changes in engine load and speed by changing their effective length in two or three stages are called _____ systems.
   A. variable inertia
   B. cross-flow
   C. intake charging
   D. either A or C

_____ 5. The method of warming the incoming air by passing it through a shroud around the exhaust manifold is called a(n) _____ system.
   A. early fuel evaporation
   B. manifold preheating
   C. mass airflow heating
   D. heated induction

_____ 6. The volume of air entering a cylinder during intake in relation to the internal volume of the cylinder when the piston is at bottom dead center is called _____ and is usually expressed as a percentage.
   A. cylinder volume
   B. induction volume
   C. volumetric efficiency
   D. compression efficiency

_____ 7. _____ increases air pressure in the intake manifold above atmospheric pressure.
   A. Back-pressure
   B. Forced induction
   C. Intake air heating
   D. Both A and C

_____ 8. When air pressure in the intake manifold reaches a preset level in a turbocharged engine, the _____ automatically directs the exhaust gases so they bypass the turbine.
   A. wastegate
   B. bypass valve
   C. blow-by valve
   D. aspirator

_____ 9. When manifold pressure rises above normal and works against the spinning exhaust turbine wheel, slowing it considerably, it is called _____.
   A. back-pressure
   B. blow-back
   C. forced induction
   D. compressor surge

_____ 10. When the outgoing pulse from one cylinder is timed to arrive at the header junction at exactly the right time to help draw out the pulse from another cylinder, it is known as a _____ exhaust.
   A. balanced
   B. cross-flow
   C. tuned
   D. mandrel-formed

_____ 11. The _____ is attached to the exhaust manifold and connects to the catalytic converter.
   A. engine pipe
   B. flexible connector
   C. down pipe
   D. either A or C

_____ **12.** The _____ helps with the alignment of the exhaust pipes as the engine moves under load.
    A. engine pipe
    B. flexible connector
    C. exhaust manifold
    D. intermediate pipe

_____ **13.** A _____ catalytic converter converts oxides of nitrogen back into nitrogen and oxygen, and the hydrocarbons and carbon monoxide to water and carbon dioxide.
    A. two-way
    B. three-way
    C. four-way
    D. all of the above

_____ **14.** The _____ connects the catalytic converter to the muffler.
    A. flexible connector
    B. exhaust bracket
    C. intermediate pipe
    D. engine pipe

_____ **15.** A(n) _____ is sometimes used between the muffler and the exhaust outlet to reduce any resonance levels that the muffler could not adequately suppress.
    A. baffle chamber
    B. intermediate pipe
    C. resonator
    D. secondary muffler

True/False

If you believe the statement to be more true than false, write the letter "T" in the space provided. If you believe the statement to be more false than true, write the letter "F".

_____ **1.** Vaporization starts when the fuel system atomizes the fuel by breaking it up into very small particles by spraying it into the charge of air.

_____ **2.** The throttle body controls airflow with a butterfly valve or valves, also called throttle valves.

_____ **3.** In many vehicles, the air cleaner is mounted where it can obtain cool, clean air.

_____ **4.** Intake manifolds were originally made from light alloy metal castings and later from cast iron.

_____ **5.** Cylinder heads that have intake and exhaust manifolds on opposite sides of the engine are known as cross-flow heads.

_____ **6.** The air intake manifold for an electronic fuel injection multipoint engine normally has short branches of variable length.

_____ **7.** Port fuel-injected or gasoline direct-injected intake manifolds do not normally need to be heated, since the manifold does not carry fuel.

_____ **8.** Electric heaters can be placed between the intake manifold and the throttle body and used to preheat the air.

_____ **9.** In a stock naturally aspirated engine, one without forced induction, volumetric efficiency can almost never be 100%.

_____ **10.** In a turbocharged engine a bypass valve may be installed so if the wastegate should fail, it can prevent an abnormal rise in manifold pressure.

_____ **11.** On many current vehicles, the exhaust manifold is often replaced with a header.

_____ **12.** Leaded fuel must not be used in an engine with a catalytic converter, because lead will coat the catalyst and prevent it from doing its job.

_____ **13.** Noise cancellation refers to putting a sound material around a perforated pipe that the exhaust gases flow through.

_____ **14.** Adding a variable-flow exhaust to the baffle or chamber system reduces emission noise.

_____ **15.** Exhaust gaskets can be found between the engine cylinder head and the exhaust manifold, the engine pipe and the catalytic converter, and possibly the catalytic converter and the muffler.

Fill in the Blank

Read each item carefully, and then complete the statement by filling in the missing word(s).

1. The _____ system ensures that clean, dry air is supplied to the engine.

2. The _____ system provides a path for the burned exhaust gases to safely exit the engine and travel out the rear of the vehicle.

3. The _____ _____ filters the incoming air.

4. A(n) _____ _____ sensor can measure air entering the engine down to tenths of a gram per second.

5. A _____ resonator is a container that is sealed and specially shaped to cancel noise created by pressure waves.

6. The _____ _____ has several tubular branches and carries air and/or air/fuel mixture from the air cleaner to the cylinder head.

7. On vehicles with emission controls, air cleaners use a(n) _____ valve to control how much hot air enters the air cleaner.

8. A buildup of pressure in the system that interferes with the outward flow of exhaust gases is known as _____-_____.

9. A(n) _____ compresses the air in the intake system to above atmospheric pressure, which increases the density of the air entering the engine.

10. A(n) _____ is a forced induction system that uses wasted kinetic energy from the exhaust gases to increase the intake pressure.

11. The purpose of a(n) _____ is to reduce the intake air temperature up to a few hundred degrees Fahrenheit before it enters the intake manifold.

12. Headers provide a(n) _____ effect to help remove exhaust gases from the cylinders.

13. There may be a flexible connection between the _____ _____ and an intermediate pipe.

14. The exhaust components are supported by _____-mounted exhaust brackets that prevent vibrations from being felt by the driver.

15. The function of a vehicle's _____ is to minimize the sounds coming from the exhaust system.

16. Noise _____ is a system that prevents the sound waves from leaving the exhaust system by canceling them out inside the muffler.

17. The _____ type muffler uses a perforated tube or baffle that is wrapped in fiberglass material, and this will absorb the noise of combustion as gases flow past.

18. A(n) _____ muffler is designed to produce anti-noise without restricting exhaust flow.

19. The _____ takes the exhaust gases away from the vehicle and must not allow any of the exhaust gases to enter the vehicle.

20. _____ _____ is a flammable material.

Labeling

Label the following diagrams with the correct terms.

1. Air induction system:

A. _____

B. _____

C. _____

D. _____

E. _____

F. _____

G. _____

H. _____

I. _____

J. _____

2. Using hot engine coolant to heat the manifold:

A. _____

B. _____

C. _____

D. _____

E. _____

3. Volumetric efficiency of a normally aspirated engine:

A. _____

B. _____

C. _____

D. _____

4. Volumetric efficiency of a forced-induction engine:

A. _____

B. _____

C. _____

D. _____

5. Identify the following exhaust system components:

A. _____

B. _____

C. _____

D. _____

E. _____

F. _____

G. _____

Skill Drills

Test your knowledge of skill drills by filling in the correct words in the photo captions.

    1. Inspecting the Throttle Body, Air Induction System, and Intake Manifold:

Step 1: Start the _____, and let the _____ stabilize. Using an _____ or _____ testing tool, place the hose near the suspected _____ area.

Step 2: Open the _____ _____ to slowly _____ the acetylene or propane.

Step 3: On vehicles _____ an idle control system, observe the rpm and smoothness of the engine. If the engine speed _____ or smooths out, then a _____ leak is present. Determine any necessary repairs. On vehicles with an _____ _____ system, connect a scan tool and select _____ _____. Observe the oxygen, short-term fuel trims, and _____ _____ width as you are _____ the acetylene or propane around.

Step 4: Shut off the _____, and connect a _____ _____ to the _____ manifold. Start the smoke machine, and _____ smoke into the intake manifold.

Step 5: Using a bright _____, look around the engine compartment for _____ of _____. Any smoke coming from the _____ _____ inlet is normal. Determine any necessary action(s).

2. Inspecting the Exhaust System for Leaks:

Step 1: _____ the vehicle on a _____, and secure it in place. Use a _____ _____ to inspect for leaks or _____ holes in the exhaust system. Check the exhaust _____ for evidence of exhaust leaks.

Step 2: Have a helper hold a _____ against the exhaust pipe(s) while the engine is _____ to increase the _____ in the system. _____ and _____ for any leaks.

Step 3: Use large adjustable _____ to test the _____ of the pipes by moderately _____ the pipes. Determine necessary repairs.

Crossword Puzzle

Use the clues to complete the puzzle.

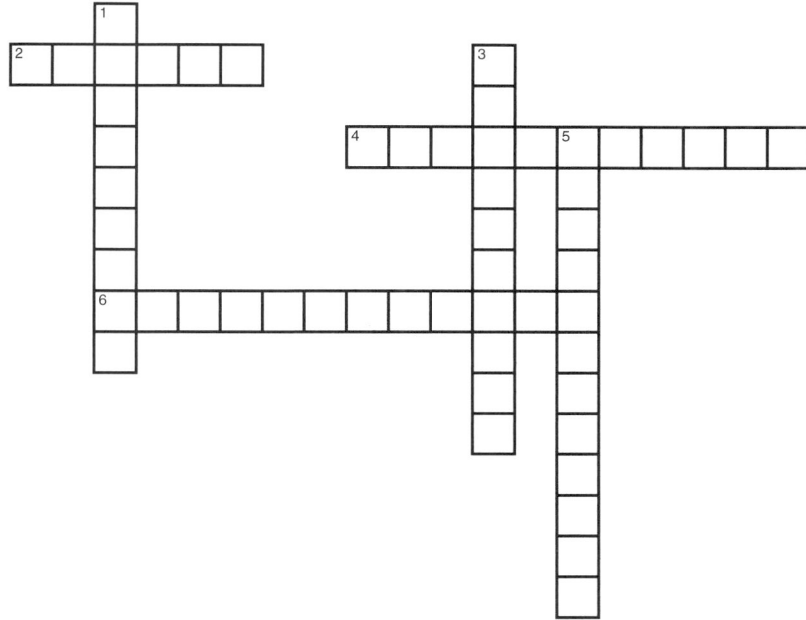

Across

2. A specially tuned exhaust manifold typically made of exhaust pipes. These pipes are usually made equal length to ensure equal flow between cylinders.
4. The pressure control valve of the supercharger. When this valve opens, it lets air move around the supercharger compressor section, reducing the boost pressure.
6. A device that pressurizes airflow into the engine. This device works similar to a supercharger, but is driven by the exhaust gases.

Down

1. A pressure regulator device that allows control of the pressure produced by the turbocharger. The device moves to allow exhaust gases to bypass the turbine wheel of the turbocharger and flow down the exhaust pipe.
3. The process of exhaust flow pulling fresh air into the cylinder as it creates a low-pressure area in the cylinder as it moves out of the cylinder.
5. A device that pressurizes the airflow into the engine, working similar to a turbocharger. This device is driven by the crankshaft through a belt or gears and does not require power from the engine.

ASE-Type Questions

Read each item carefully, and then select the best response.

_____ 1. Tech A says that coolant circulates through some intake manifolds to help warm them up. Tech B says that some intake manifolds use an electric heater grid to warm up the intake air. Who is correct?
   A. Tech A
   B. Tech B
   C. Both A and B
   D. Neither A nor B

_____ **2.** Tech A says that in some gas engines, the intake manifold runner length can be changed. Tech B says that some intake systems use resonators to quiet intake noise. Who is correct?
  A. Tech A
  B. Tech B
  C. Both A and B
  D. Neither A nor B

_____ **3.** Tech A says that a good way to check the integrity of exhaust pipes is tap on them with the handle of a screw driver. Tech B says that small exhaust leaks can be found with a smoke machine. Who is correct?
  A. Tech A
  B. Tech B
  C. Both A and B
  D. Neither A nor B

_____ **4.** Tech A says that scavenging creates a low pressure in the cylinder by means of the exhaust gases flowing through the exhaust pipe. Tech B says that back-pressure in the exhaust system increases the scavenging effect. Who is correct?
  A. Tech A
  B. Tech B
  C. Both A and B
  D. Neither A nor B

_____ **5.** Tech A says that shorter intake manifolds produce higher torque at lower rpm. Tech B says that longer intake manifolds produce higher torque at high engine speeds. Who is correct?
  A. Tech A
  B. Tech B
  C. Both A and B
  D. Neither A nor B

_____ **6.** Tech A says that air becomes heated as a supercharger compresses it. Tech B says that a turbocharger pulls the exhaust gases out of the engine, thereby increasing the scavenging effect. Who is correct?
  A. Tech A
  B. Tech B
  C. Both A and B
  D. Neither A nor B

_____ **7.** Tech A says that a wastegate is designed to direct waste gases to the intake manifold. Tech B says that a blow-off valve vents excess boost pressure from the turbocharger when the throttle is closed quickly. Who is correct?
  A. Tech A
  B. Tech B
  C. Both A and B
  D. Neither A nor B

_____ **8.** Tech A says that the exhaust system is usually suspended from the underbody by rubber mounts. Tech B says some engines use a flexible pipe between the engine pipe and intermediate pipe. Who is correct?
  A. Tech A
  B. Tech B
  C. Both A and B
  D. Neither A nor B

_____ **9.** Tech A says that a catalytic converter operates best at cold temperatures. Tech B says that the catalytic converter is typically located after the muffler. Who is correct?
  A. Tech A
  B. Tech B
  C. Both A and B
  D. Neither A nor B

_____ **10.** Tech A says that small vacuum leaks can be found with a vacuum gauge. Tech B says that the catalytic converter can be ruined if it becomes contaminated with lead or silicone. Who is correct?
  A. Tech A
  B. Tech B
  C. Both A and B
  D. Neither A nor B

Emission Control

Chapter Review

The following activities have been designed to help you refresh your knowledge of this chapter. Your instructor may require you to complete some or all of these activities as a regular part of your training program. You are encouraged to complete any activity that your instructor does not assign as a way to enhance your learning.

Matching

Match the following terms with the correct description or example.

A. Air diverter valve

B. Blow-by

C. Carbon monoxide

D. Compression-ignition engine

E. Emission

F. Exhaust gas recirculation (EGR) system

G. Flame front

H. Hydrocarbon

I. Lambda

J. Particulate matter

K. PCV valve

L. Photochemical smog

M. Positive crankcase ventilation (PCV) system

N. Purge switch

O. Scavenging effect

P. Stoichiometric ratio

Q. Stroke

R. Two-way catalytic converter

S. Vaporization

T. Vent solenoid

_____ **1.** The exact ratio between air and fuel, at which both are burned completely.

_____ **2.** A valve that controls the amount of crankcase ventilation flow that is allowed and varies with changes in manifold pressure.

_____ **3.** The rapid burning of the air/fuel mixture that moves outward from the spark plug across the cylinder.

_____ **4.** Pressure that leaks past the compression rings during compression and combustion.

_____ **5.** A condition caused by moving columns of air, which create a low-pressure area behind them, resulting in a pulling force that is used to pull the remaining burned gases from the combustion chamber.

_____ **6.** A system that draws blow-by gases from the crankcase into the intake to be burned.

_____ **7.** A converter that changes only hydrocarbons and carbon monoxide into harmless elements.

_____ **8.** A system that recirculates a portion of burned gases back into the combustion chamber to displace air and fuel and cool combustion temperatures.

_____ **9.** A solenoid that allows fresh air to enter the evaporative system during a purge event. Also used for an evaporative system monitoring test.

_____ **10.** A molecule made up of hydrogen and carbon. It is considered a harmful vehicle emission.

_____ **11.** An engine that uses the heat of compression to ignite the air/fuel mixture; also known as a diesel engine.

_____ **12.** Particles that are heavier than air that are released during combustion. It is the black smoke that is commonly emitted from a diesel engine.

_____ **13.** A valve that changes secondary airflow from the exhaust stream to the air cleaner or vice versa.

_____ **14.** A brown haze that hangs in the sky, typically seen over large cities. Smog is a major health issue to humans because it affects lung tissue.

_____ **15.** The ability of a liquid to evaporate.

_____ **16.** A poisonous gas released during combustion of rich air/fuel mixtures.

_____ **17.** A device used to show the computer when purge is occurring. It is used as feedback device to allow the computer to determine whether flow is happening.

_____ **18.** A gas that is released to the atmosphere; usually refers to a harmful gas.

_____ **19.** The movement of the piston in the engine from top dead center to bottom dead center, or vice versa.

_____ **20.** The ratio of air to fuel at which all of the oxygen in the air and all of the fuel are completely burned.

Multiple Choice

Read each item carefully, and then select the best response.

_____ **1.** The _____ standard is an automotive self-diagnostic system mandated by the EPA that requires a warning light to alert the driver of an emission system fault.
   A. clean air
   B. emission control
   C. OBDI
   D. OBDII

_____ **2.** Oxides of nitrogen are categorized as _____ emission gases.
   A. nonharmful
   B. harmful
   C. debatable
   D. inert

_____ **3.** _____ is a by-product of complete combustion.
   A. Water
   B. Nitrogen
   C. Carbon dioxide
   D. Both A and C

_____ **4.** Since carbon monoxide is absorbed so easily by red blood cells, the _____ has set a permissible exposure limit in the workplace of 35 ppm for an 8-hour shift with a ceiling of 200 ppm for any length of time.
   A. Occupational Safety and Health Administration
   B. National Institute for Occupational Safety and Health
   C. American Conference of Governmental Industrial Hygienists
   D. Society of Automotive Engineers

_____ **5.** Which corrosive compound, emitted into the atmosphere through exhaust, is a major environmental pollutant, coming back to earth in contaminated rainwater, called acid rain?
   A. Oxides of nitrogen
   B. Carbon monoxide
   C. Sulfur dioxide
   D. Hydrocarbons

_____ **6.** What is produced in the atmosphere when unburned hydrocarbons and oxides of nitrogen react chemically with sunlight?
   A. Smog
   B. Acid rain
   C. Sulfur dioxide
   D. Carbon monoxide

_____ **7.** The stoichiometric ratio for compressed natural gas is _____.
   A. 14.7:1
   B. 17.2:1
   C. 6.45:1
   D. 34.3:1

_____ **8.** The _____ is an example of a precombustion emission control system.
   A. exhaust gas recirculation system
   B. secondary air injection system
   C. heated air intake system
   D. both A and C

_____ 9. _____ catalytic converters contain both a reduction catalyst and an oxidizing catalyst.
   A. Two-way
   B. Three-way
   C. Four-way
   D. All of the above

_____ 10. The _____ is a type of positive crankcase ventilation system.
   A. fixed orifice
   B. variable orifice
   C. separator
   D. all of the above

_____ 11. The _____ system is designed to ensure hydrocarbons are not released into the atmosphere when fuel in the fuel tank begins to vaporize and build pressure.
   A. evaporative emission
   B. positive crankcase ventilation
   C. exhaust gas recirculation
   D. secondary air injection

_____ 12. Secondary air injection systems use a(n) _____ to control airflow in the system.
   A. air-switching valve
   B. air diverter valve
   C. check valve
   D. all of the above

True/False

If you believe the statement to be more true than false, write the letter "T" in the space provided. If you believe the statement to be more false than true, write the letter "F".

_____ 1. Humans breathe in oxygen and exhale carbon dioxide; trees and plants take in carbon dioxide and give back oxygen.

_____ 2. The EPA requires that when a vehicle's emissions deviate from the federal test procedure limit by 3.5 times, the malfunction indicator lamp must turn on and one or more specific diagnostic trouble codes be set in memory.

_____ 3. Particulate matter is graded in a size range from 10 nanometers to 100 micrometers in diameter; particulates of less than 10 micrometers are dangerous to humans.

_____ 4. Gasoline, diesel, LPG, and natural gas are all hydrocarbon compounds.

_____ 5. Exposure to carbon monoxide levels of 400 ppm may be fatal in as little as 3 hours, and 6400 ppm may be fatal in as little as 30 minutes.

_____ 6. Sulfur reduces catalyst efficiency in modern vehicles, and vehicles operating with higher sulfur gasoline have higher emissions than vehicles operating on lower sulfur gasoline.

_____ 7. Keeping the air/fuel ratio close to the stoichiometric point produces minimal hydrocarbon and carbon monoxide emissions.

_____ 8. The exhaust gas recirculation system is a postcombustion emission control system.

_____ 9. Any vehicle produced after 1996 monitors catalyst efficiency by using two oxygen sensors, one in front of the catalytic converter and one in the exhaust stream.

_____ 10. The positive crankcase ventilation system regulates the flow of blow-by gases between the crankcase and the intake manifold.

_____ 11. A PCV valve connects the exhaust port, or manifold, and the intake manifold.

_____ 12. In the combustion chamber, lower combustion temperatures tend to create more power.

_____ 13. The EVAP system stores the fuel vapors in the charcoal canister until the ECM determines that they can be burned in the engine without affecting the drivability of the vehicle.

_____ 14. Some manufacturers use dedicated sensors such as a purge switch, which is located in the purge line and senses the flow of gases or the pressure drop caused by the flow of gases that are being purged.

_____ 15. The air diverter valve is used to switch air from upstream to downstream once the engine reaches the specified temperature.

Fill in the Blank

Read each item carefully, and then complete the statement by filling in the missing word(s).

1. _____ _____ systems are designed to limit the pollution caused by the storing and burning of various fuels.

2. The _____ _____ is the maximum amount of emissions that a vehicle is permitted to emit.

3. The _____ _____ completes the combustion process by adding another oxygen molecule to the carbon monoxide to convert it to carbon dioxide.

4. Victims of _____ _____ poisoning can sometimes look healthy and pink-cheeked because concentrations of carbon monoxide in the bloodstream give the blood a brighter red color than normal.

5. _____ _____ _____ are claimed to be major contributors to photochemical smog, along with hydrocarbons and sunlight.

6. In spark-ignition engines, _____ are caused by incomplete combustion of rich air/fuel mixtures.

7. In a combustion chamber where surface temperatures are low, the combustion flame can go out or be _____.

8. Reducing the valve overlap reduces the _____ _____, which is when the exhaust gases moving out of the combustion chamber create a low-pressure area in the exhaust manifold, helping to pull air and fuel into the cylinder.

9. The _____ _____, also referred to as lambda, is the ratio of air to fuel at which all of the oxygen in the air and all of the fuel are completely burned.

10. The _____ emitted from the internal combustion can be reduced by taking actions both prior to combustion and after combustion.

11. In older vehicles, crankcase vapors were vented directly to the atmosphere through a(n) _____ _____ or road-draft tube.

12. The _____ _____ _____ system was designed by automotive engineers in the 1970s to control the emission of oxides of nitrogen.

13. OBDII made monitoring of the EGR system mandatory; each manufacturer became responsible for ensuring its system runs a self-test called a(n) EGR _____ _____ test to verify that it is operating correctly.

14. Newer EVAP systems monitor purge operation by reading a _____ _____ pressure sensor.

15. The _____ _____ intake system collects hot air from around the exhaust manifold and mixes it with outside air entering the air cleaner assembly.

Labeling

Label the following diagrams with the correct terms.

1. PCV system:

A. _____

B. _____

C. _____

D. _____

2. Computer-controlled EGR system:

A. _____

B. _____

C. _____

D. _____

E. _____

F. _____

G. _____

H. _____

3. EVAP system:

A. _____

B. _____

C. _____

D. _____

E. _____

F. _____

G. _____

H. _____

I. _____

4. Air diverter valve and check valve:

A. _____

B. _____

C. _____

D. _____

E. _____

5. Heated air intake system:

A. _____

B. _____

C. _____

D. _____

E. _____

Skill Drills

Test your knowledge of skill drills by filling in the correct words in the photo captions.

1. Inspecting and Servicing the PCV System:

Step 1: Locate the _____ _____. With the engine idling, _____ the PCV _____ and check for the presence of a strong _____.

Step 2: Remove the _____ and check that it is still pliable and not _____ with _____ deposits.

Step 3: Remove the PCV valve and _____ it for _____. If it is _____, is sludged up, or sticks, _____ it with a new one of the same type.

Step 4: Reinstall the PCV valve into the _____ _____, and remove the _____ hose from the _____ _____ assembly.

Step 5: With the engine _____, block off the breather hose and _____ for vacuum _____ up in the breather hose and _____, indicating that the PCV system can handle the amount of _____-_____ gases.

Crossword Puzzle

Use the clues to complete the puzzle.

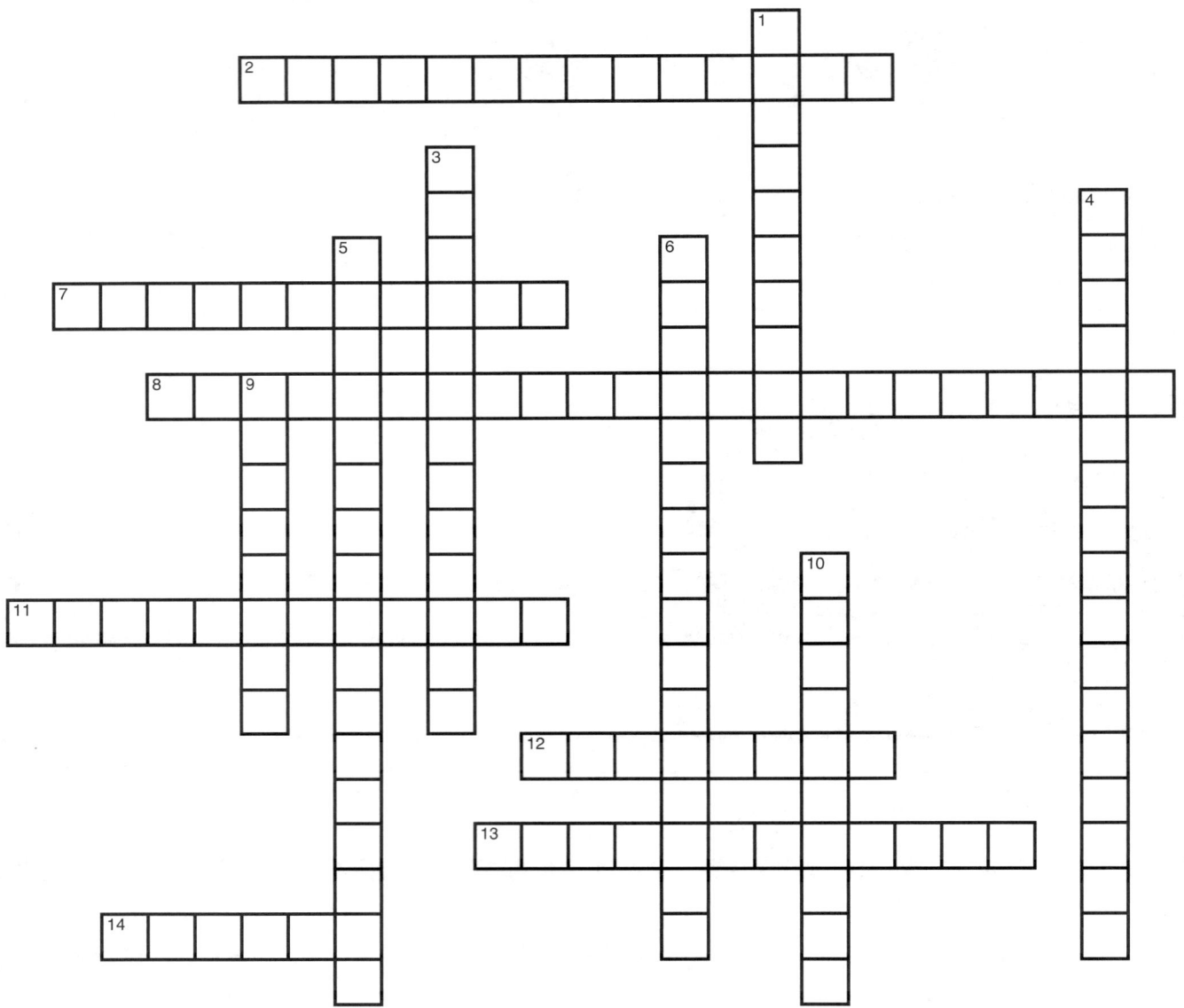

Across

2. A poisonous gas released during combustion of rich air-fuel mixtures.
7. A molecule made up of hydrogen and carbon. It is considered a harmful vehicle emission.
8. A sensitive pressure sensor mounted in the fuel tank or EVAP system used to monitor the system for leaks.
11. The ability of a liquid to evaporate.
12. The state in a combustion chamber in which the flame cannot burn due to cold surfaces or poor distribution of the fuel mixture.
13. A tube used in the PCV system to allow fresh air into the engine crankcase.
14. The movement of the piston in the engine from top dead center to bottom dead center, or vice versa.

Down

1. The maximum amount of emissions that a vehicle is permitted to emit. Values are assigned to different classifications of vehicles.
3. A by-product of complete combustion, carbon dioxide is a chemical compound made up of two oxygen molecules bonded to a single carbon molecule.
4. A brown haze that hangs in the sky; typically over large cities. Smog is a major health issue to humans because it affects lung tissue.
5. Particles that are heavier than air that are released during combustion. It is the black smoke that is commonly emitted from a diesel engine.
6. A control module diagnostic test that is run to ensure that the EGR valve and passageways are operating properly.
9. A gas that is released to the atmosphere; usually refers to a harmful gas.
10. The rapid burning of the air-fuel mixture that moves outward from the spark plug across the cylinder.

ASE-Type Questions

Read each item carefully, and then select the best response.

_____ **1.** Tech A says that the purge valve is part of the evaporative emission system. Tech B says that OBDII systems must illuminate the MIL when the emissions exceed 1.5 times the FTP. Who is correct?
   A. Tech A
   B. Tech B
   C. Both A and B
   D. Neither A nor B

_____ **2.** Tech A says that carbon monoxide is partially burned fuel. Tech B says that oxides of nitrogen are unburned fuel. Who is correct?
   A. Tech A
   B. Tech B
   C. Both A and B
   D. Neither A nor B

_____ **3.** Tech A says that hydrocarbons are a result of complete combustion. Tech B says that a catalytic converter creates a chemical reaction, changing carbon monoxide and hydrocarbons to water and carbon dioxide. Who is correct?
   A. Tech A
   B. Tech B
   C. Both A and B
   D. Neither A nor B

_____ **4.** Tech A says that allowing hot exhaust gases into the engine through the EGR valve helps to warm up the engine during warm-up. Tech B says that rich fuel mixtures will create low amounts of carbon monoxide. Who is correct?
   A. Tech A
   B. Tech B
   C. Both A and B
   D. Neither A nor B

_____ **5.** Tech A says that burning gasoline in an engine creates water as a by-product of combustion. Tech B says that variable valve timing can be used to reduce oxides of nitrogen, eliminating the need for an EGR valve on some engines. Who is correct?
   A. Tech A
   B. Tech B
   C. Both A and B
   D. Neither A nor B

_____ **6.** Tech A says that the PCV system recirculates exhaust gases into the intake manifold. Tech B says that the PCV system recirculates blow-by gases into the intake manifold. Who is correct?
  A. Tech A
  B. Tech B
  C. Both A and B
  D. Neither A nor B

_____ **7.** Tech A says that variable orifice PCV valves allow maximum PCV flow at idle. Tech B says that lean means there is not enough air in the air-fuel mixture. Who is correct?
  A. Tech A
  B. Tech B
  C. Both A and B
  D. Neither A nor B

_____ **8.** Tech A says that the EVAP system stores escaping hydrocarbons until the engine can burn them. Tech B says that catalytic converters were designed for regular leaded gas. Who is correct?
  A. Tech A
  B. Tech B
  C. Both A and B
  D. Neither A nor B

_____ **9.** Tech A says that the EGR valve is open fully at idle so that the engine will not die. Tech B says that oxides of nitrogen are created in large amounts when the combustion temperature is above 2500°F (1400°C). Who is correct?
  A. Tech A
  B. Tech B
  C. Both A and B
  D. Neither A nor B

_____ **10.** Two technicians are discussing PCV system failures. Tech A says that the hose to the PCV valve can crack and cause a large vacuum leak. Tech B says that the hose to the PCV valve can become plugged with sludge restricting PCV flow. Who is correct?
  A. Tech A
  B. Tech B
  C. Both A and B
  D. Neither A nor B

Electric and Hybrid Vehicles

Workbook Activities

The following activities have been designed to help you. Your instructor may require you to complete some or all of these activities as a regular part of your *Fundamentals of Automotive Technology training program*. You are encouraged to complete any activity that your instructor does not assign as a way to enhance your learning in the classroom.

Chapter Review

The following exercises provide an opportunity to refresh your knowledge of this chapter.

Matching

Match the following terms with the correct description or example.

A. Alternative fuel
B. Battery electric vehicle (BEV)
C. Continuously variable transmission
D. Displacement-on-demand
E. Electric machine
F. Hybrid electric vehicle (HEV)
G. Idle stop
H. Inverter
I. Multiphase
J. Parallel hybrid

K. Permanent magnet
L. Plug-in hybrid electric vehicle (PHEV)
M. Power divider/splitter
N. Power-on-demand
O. Regenerative braking
P. Retarding effect
Q. Series hybrid
R. Series-parallel hybrid
S. Stator winding
T. Volumetric efficiency

_____ 1. A hybrid electric vehicle that uses the internal combustion engine and/or the battery pack for propulsion.

_____ 2. A vehicle powered by battery only.

_____ 3. A feature that shuts down the engine when not needed to save fuel.

_____ 4. A transmission without individual gears or gear ratios.

_____ 5. A hybrid electric vehicle powered by an internal combustion engine, but driven by battery

_____ 6. Another name for a traction motor with regenerative capability.

_____ 7. A feature that turns off the internal combustion engine when the vehicle is at a standstill.

_____ 8. A winding in a motor that creates the needed magnetism for the motor to rotate.

_____ 9. A device that converts direct current to alternating current.

_____ 10. A hybrid vehicle driven simultaneously by an internal combustion engine and an electric machine.

_____ 11. An electric motor that operates through more than one phase.

_____ 12. A feature that allows drive motors to act as generators to recharge the traction batteries during deceleration or when braking.

_____ 13. A hybrid electric vehicle in which only one power source, the battery, is used to propel the vehicle for a certain distance, limited by the storage capacity of the battery and the efficiency of the motor.

_____ 14. A nonpetroleum-based motor fuel.

_____ **15.** A material with natural or manmade constant magnetic properties.

_____ **16.** A device that receives power from an internal combustion engine and electric machine to power a hybrid electric vehicle.

_____ **17.** The result of slowing the vehicle.

_____ **18.** A vehicle that uses two power sources for propulsion, one of which is electricity.

_____ **19.** A measure of how well an internal combustion engine can breathe at a given rpm at wide-open throttle.

_____ **20.** A feature that allows cylinders to be taken off-line when not needed, such as at vehicle cruise.

Multiple Choice

Read each item carefully, and then select the best response.

_____ **1.** What type of electric motor does not use brushes or commutators?
  A. Permanent magnet
  B. Multiphase
  C. Brushless
  D. Synchronous

_____ **2.** In a _____ electric motor the stator contains more than one winding, usually three.
  A. Synchronous
  B. Brushless
  C. Multiphase
  D. Permanent magnet

_____ **3.** What type of vehicle uses a combination of electric power and an internal combustion engine?
  A. Fuel cell vehicle
  B. Hybrid electric vehicle
  C. Battery electric vehicle
  D. Flexible-fuel vehicle

_____ **4.** In a(n) _____ hybrid drive configuration the gasoline engine used only to drive a generator and charge a battery, which produces power to drive an electric motor.
  A. Series hybrid
  B. Parallel hybrid
  C. Series-parallel hybrid
  D. All of the above

_____ **5.** In a hybrid vehicle the _____ feature temporarily shuts off the internal combustion engine when it is not needed, such as when idling or coasting.
  A. Regenerative braking
  B. Displacement-on-demand
  C. Idle stop
  D. Power-on-demand

_____ **6.** The _____ feature turns off the internal combustion engine when the vehicle is at a standstill.
  A. Idle stop
  B. Regenerative braking
  C. Power-on-demand
  D. Displacement-on-demand

_____ **7.** A power _____ is used to change direct current from the battery into alternating current to power the electric motors.
  A. Converter
  B. Inverter
  C. Rectifier
  D. Coupling

_____ **8.** What type of vehicle uses only one power source, the battery, to propel the vehicle for a certain distance before the batteries need charging and the internal combustion engine is needed?
   A. Plug-in hybrid electric vehicle
   B. Fuel cell vehicle
   C. Series-parallel hybrid vehicle
   D. Flexible-fuel vehicle

_____ **9.** The high voltage wiring used in hybrid vehicles is usually _____ in color.
   A. Red
   B. Black
   C. Orange
   D. Green

_____ **10.** On most hybrid vehicles, the _____ auxiliary battery is used to power up the vehicles computers and run virtually all of the accessories.
   A. 120-volt
   B. 24-volt
   C. 12-volt
   D. 9-volt

True/False

If you believe the statement to be more true than false, write the letter T in the space provided. If you believe the statement to be more false than true, write the letter F.

_____ **1.** The AC systems used in on-road vehicles operate on dangerously high voltages, some as high as 500 to 650 volts.

_____ **2.** Electric vehicles for road use today use DC motors.

_____ **3.** US light-duty hybrids are almost exclusively gasoline-electric.

_____ **4.** A parallel hybrid allows both the engine and the electric motor to drive the vehicle together, or either one by itself.

_____ **5.** The parallel hybrid arrangement is best suited to vehicles that travel greater distances between stops.

_____ **6.** Displacement-on-demand deactivates cylinders when operating under cruise or coasting conditions to save fuel.

_____ **7.** The power-on-demand feature deactivates cylinders when operating under cruise or coasting conditions to save fuel.

_____ **8.** The motor control unit controls the power divider/splitter so that the drive remains at its most efficient.

_____ **9.** Plug-in hybrid electric vehicles have a relatively small auxiliary ICE to drive a generator for charging the batteries.

_____ **10.** Fuel cell vehicles are considered hybrid vehicles because they have two power sources.

_____ **11.** The Chevy Volt is an example of a fuel cell vehicle.

Fill in the Blank

Read each item carefully, and then complete the statement by filling in the missing word(s).

1. A(n) _____ fuel is essentially anything other than a petroleum-based motor fuel (gasoline or diesel fuel) that is used to propel a motorized vehicle.

2. The electric motor in some power trains is also used as a(n) _____ to provide partial recharging of batteries when the vehicle is coasting.

3. In a synchronous electric motor the speed of the rotor is synchronized to the frequency of current flowing through the _____ windings.

4. When coupled with sustainable or renewable energy sources, the _____ _____ vehicle can truly serve as an emissions-free vehicle.

5. Solar panels, wind generators, and geothermal sources are all examples of _____ energy sources.

6. _____ braking occurs when the drive motor(s) act as generators to recharge the traction batteries during deceleration or when braking.

7. The continuously _____ transmission does not "shift" from one gear to another in the usual sense; rather, it continuously changes the amounts of speed and torque from the engine and the motor to the wheels.

8. The _____ effect (vehicle slowing) caused by regenerative braking provides the same deceleration as normal engine braking would.

9. The main high-voltage battery pack must be _____ before service is performed on or around the high-voltage components.

10. If you are not properly _____ to work on a particular hybrid vehicle, you should not work on it.

Labeling

Label the following diagrams with the correct terms.

1. Power splitting transmission:

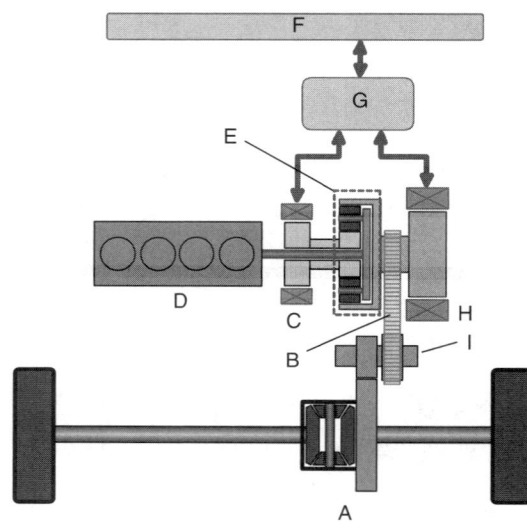

A. _____

B. _____

C. _____

D. _____

E. _____

F. _____

G. _____

H. _____

I. _____

Crossword Puzzle

Use the clues in the column to complete the puzzle.

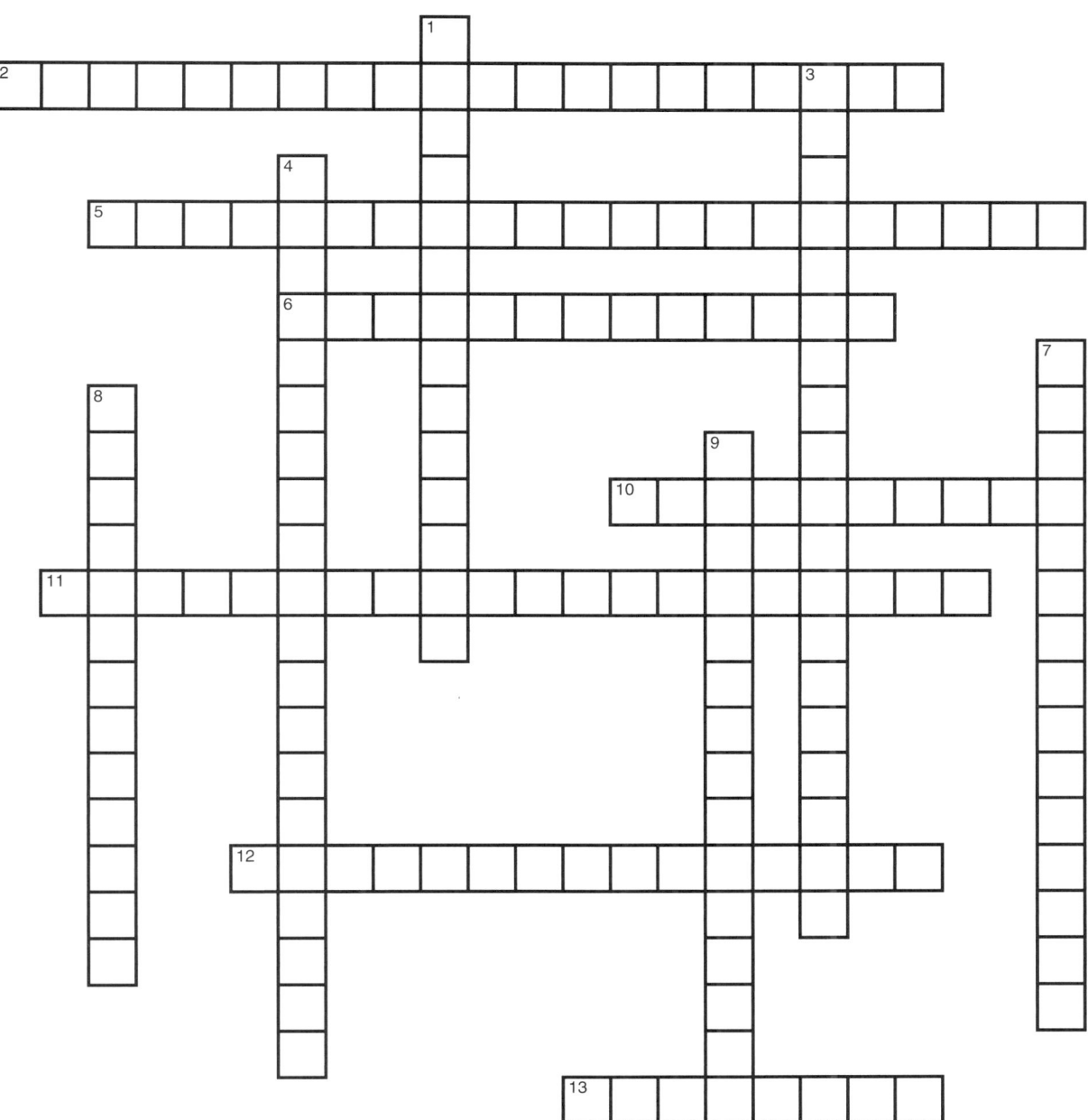

Across

2. A hybrid electric vehicle that uses internal combustion engine and/or the battery pack for propulsion.

5. A vehicle that uses two power sources for propulsion, one of which is electricity.

6. A device that receives power from an internal combustion engine and electric machine to power a hybrid electric vehicle.

10. An electric motor that operates through more than one phase.

11. A measure of how well an internal combustion engine can breathe at a given rpm at wide-open throttle.

12. A material with natural or man-made constant magnetic properties.

13. A feature that turns off the internal combustion engine when the vehicle is at a standstill.

Down

1. A hybrid vehicle driven simultaneously by an internal combustion engine and an electric machine.

3. A feature that allows drive motors to act as generators to recharge the traction batteries during deceleration or when braking.

4. A feature that allows cylinders to be taken off-line when not needed, such as at vehicle cruise.

7. A nonpetroleum-based motor fuel.

8. A winding in an alternator that creates the current output, or a winding in a motor that creates the needed magnetism for the motor to rotate.

9. Another name for a traction motor with regenerative capability.

ASE-Type Questions

Read each item carefully, and then select the best response.

_____ 1. Tech A says that alternative fuels may be liquid or gas (or electric) and contain a variety of latent heat energy for use in the internal combustion engine. Tech B says that an alternative fuel vehicle is a vehicle powered by something other than petroleum. Who is correct?
   A. Tech A
   B. Tech B
   C. Both A and B
   D. Neither A nor B

_____ 2. Tech A says that battery electric vehicles (BEVs) were introduced more than a century ago. Tech B says that BEVs have only been around for approximately the past 10 years. Who is correct?
   A. Tech A
   B. Tech B
   C. Both A and B
   D. Neither A nor B

_____ 3. Tech A says that high-voltage wires in hybrid and electric vehicles are usually a special color. Tech B says that special high-voltage shoes should be worn when working on high-voltage vehicles. Who is correct?
   A. Tech A
   B. Tech B
   C. Both A and B
   D. Neither A nor B

_____ 4. Tech A says that the electric motors used in all automotive hybrid applications are direct current, or DC, motors. Tech B says that the electric motors used in all automotive hybrid applications are alternating current, or AC, motors. Who is correct?
   A. Tech A
   B. Tech B
   C. Both A and B
   D. Neither A nor B

_____ 5. Tech A says that in a hybrid vehicle the engine usually drives a generator to provide the electricity, which is stored in batteries that then drive an electric motor. Tech B says that the engine may also drive the wheels directly as well. Who is correct?
   A. Tech A
   B. Tech B
   C. Both A and B
   D. Neither A nor B

_____ **6.** Tech A says that in hybrid vehicles the main high voltage battery pack must be disconnected before service is performed on or around the high-voltage components. Tech B says that this is not necessary because the system has built in protection devices. Who is correct?

    A. Tech A

    B. Tech B

    C. Both A and B

    D. Neither A nor B

_____ **7.** Tech A says that on most hybrid vehicles, a 12-volt auxiliary battery is used to power up the vehicles computers and run most of the accessories. Tech B says that the 12-volt auxiliary battery is typically used to crank the ICE power unit when starting. Who is correct?

    A. Tech A

    B. Tech B

    C. Both A and B

    D. Neither A nor B

_____ **8.** Tech A says that the 12V battery in a hybrid vehicle is generally smaller than in a standard vehicle. Tech B says that the battery is more susceptible to parasitic current draw, as they do not have as much current capacity. Who is correct?

    A. Tech A

    B. Tech B

    C. Both A and B

    D. Neither A nor B

_____ **9.** Tech A says that that hybrids have a feature that allows the traction motor's batteries to be charged when the vehicle is braking. Tech B says this is known as regenerative braking. Who is correct?

    A. Tech A

    B. Tech B

    C. Both A and B

    D. Neither A nor B

_____ **10.** Tech A says that when servicing a hybrid electric vehicle, it is OK to leave the power switch on as long as the ICE and electric motor are not operating. Tech B says when performing service on a hybrid electric vehicle, the key fob should be more than 15 feet away from the vehicle. Who is correct?

    A. Tech A

    B. Tech B

    C. Both A and B

    D. Neither A nor B